Transport and Energy Processes 2019

Held at the 2019 AIChE Annual Meeting

Orlando, Florida, USA
10 - 15 November 2019

ISBN: 978-1-7138-0531-1

Printed from e-media with permission by:

Curran Associates, Inc.
57 Morehouse Lane
Red Hook, NY 12571

Some format issues inherent in the e-media version may also appear in this print version.

AIChE
120 Wall Street, FL 23
New York, NY 10005-4020

Phone: (800) 242-4363
Fax: (203) 775-5177

www.aiche.org

Additional copies of this publication are available from:

Curran Associates, Inc.
57 Morehouse Lane
Red Hook, NY 12571 USA
Phone: 845-758-0400
Fax: 845-758-2633
Email: curran@proceedings.com
Web: www.proceedings.com

TABLE OF CONTENTS

559467 Probing the Mechanisms of Aqueous-Phase V^{2+}/V^{3+} and Ce^{3+}/Ce^{4+} Redox Reactions for Redox Flow Batteries

Energy demand continues to increase as the world's population and average quality of life continue to climb, and this demand is accompanied by increasing levels of CO_2. Utilizing renewable energy sources can mitigate the emissions of CO_2, but the intermittency of renewable systems such as solar and wind poses a challenge in matching energy supply to demand. Although pumped hydroelectric is the most common electrical energy storage technology, it has geographical limitations, necessitating new innovations in other energy storage technologies. One promising energy storage technology is redox flow batteries (RFBs). RFBs have the advantages of convective mass transport, long cycle lifetimes, and decoupled power and energy. However, the current densities and efficiencies of RFBs must be increased to drive down costs and make these batteries economical. Improvements to RFB performance can come from minimizing efficiency losses by fundamentally understanding the mechanism of the charge transfer in RFBs. Efficiency losses of these systems arise from mass transport, ohmic, and kinetic overpotentials. In this talk, we will focus on the fundamentals of what controls the kinetic overpotentials for two redox couples employed in aqueous RFBs.

The all vanadium redox flow battery $VO^{2+}/VO_2^{+}//V^{2+}/V^{3+}$ is the most developed RFB,[1] but still the kinetics of the V^{2+}/V^{3+} redox couple are incompletely understood. Here we focus on understanding first the structure of the ions in aqueous solution in a series of different electrolytes, using a combination of experimental spectroscopy and time dependent-density functional theory calculations. We then compare the structures in different electrolytes to carefully measured kinetic parameters (e.g., activation barriers, exchange current densities) on controlled surfaces to extract information about the mechanism of reaction. We observe from both steady state activity measurements and electrochemical impedance measurements that halides can act as bridging agents on carbon electrodes for the charge transfer reaction. These results are similar to reports of the influence of halides on Cr^{2+}/Cr^{3+} and other redox couples,[2,3] but have not previously been reported for vanadium RFB applications. These results lead to a better understanding of the V^{2+}/V^{3+} mechanism, as well as pathways for further improvements in kinetic performance through electrocatalyst and electrolyte design.

We also show results for a less commonly used, but particularly interesting, redox couple, Ce^{3+}/Ce^{4+}, which has a high positive potential,[4] enabling higher power densities for RFBs without requiring increases in current densities. We first investigate the role of the electrolyte on the structure of the cerium ions and show that although Ce^{3+} is solely complexed by water, Ce^{4+} is prone to complexation with the anion, which significantly impacts the redox potential of the couple. We compare the energy of Ce^{4+} complexation (determined from the change in redox potential) to our measured kinetic data to understand the mechanism of the reaction in aqueous phase on different electrocatalyst surfaces. The significant shift in the Ce^{3+}/Ce^{4+} redox potential with electrolyte[5] (not observed for V^{2+}/V^{3+}) also has an influence on the electrode structure during reaction. We explore the influence of the electrolyte on the electrocatalyst structure through the use of in situ X-ray Absorption Spectroscopy (e.g., changes in electrocatalyst oxidation state), and show how this may change the kinetics of the Ce^{3+}/Ce^{4+} reaction.

References

(1) Sum, E.; Skyllas-Kazacos, M. A Study of the V(II)/V(III) Redox Couple for Redox Flow Cell Applications. *J. Power Sources* **1985**, *15*, 179–190.

(2) Ulrich, J. J.; Anson, F. C. Ligand Bridging by Halide in the Electrochemical Oxidation of Chromium(II) at Mercury Electrodes. *Inorg. Chem.* **1969**, *8*, 195–200.

(3) Hung, N. C.; Nagy, Z. Kinetics of the Ferrous/Ferric Electrode Reaction in the Absence of Chloride

Catalysis. *J. Electrochem. Soc.* **1987**, *134*, 2215–2220.

(4) Tucker, M. C.; Weiss, A.; Weber, A. Z. Improvement and Analysis of the Hydrogen-Cerium Redox Flow Cell. *J. Power Sources* **2016**, *327*, 591–598.

(5) Morss, L. R. Yttrium, Lanthanum, and the Lanthanide Elements. In *Standard Potentials in Aqueous Solution*; Bard, A. J., Parsons, R., Jordan, J., Eds.; Marcel Dekker, Inc.: New York and Basel, 1985; pp 619–621.

560134 Scalable Faradaic Electrodes for Enhanced Charge Transfer in Redox Flow Batteries

Further development of electrode materials to enhance interfacial charge transfer interactions in redox flow batteries is necessary to diminish costs and augment electrochemical performance. Unpurified carbon nanotubes, containing their ferrocene synthesis catalysts, are activated to permit interactions between the ferrocene nanoparticles and the bulk electrolyte. These nanoparticles drastically decrease the activation and Ohmic losses of electron transfer at the electrode/electrolyte interface which results in mass transfer limitations not effecting the discharge voltage of the battery until much greater discharge currents. Carbon nanotube electrodes are activated in acidic conditions utilizing cyclic voltammetry and results in the physical fracturing of the nanotubes. Modified electrodes experience a 141% increase in power density and a 56.5% increase in energy density in coin cell configurations. Comparable peak power densities to the literature standard was attained at 176 mW/cm^2 for all-iron redox for batteries, although our system achieved this mark in a stagnant electrolyte configuration and power density outputs will be greatly elevated when pilot redox flow battery experiments have been conducted.

577014 Flexible Ceramic Membranes for High Performance Redox Flow Batteries

Ion exchange membranes (IEM) are often one of the most expensive and performance limiting components of an end-use application. In commercial redox flow batteries (RFBs) they can account for 8 – 27% of the total battery cost. High IEM production costs are largely the result of intrinsically expensive raw materials and lengthy manufacturing processes that require tight environmental controls. In addition to high costs, many IEMs have low ion conductivity, poor ion selectivity and inadequate chemical stability that prevent RFBs from reaching their full potential. In this talk, we discuss the design and scale-up of a novel, nanoporous ceramic RFB membrane that is chemically robust, more mechanically durable, ion selective and 10x less expensive to produce than traditional IEMs.

In this work, we investigated the use of a primarily inorganic membrane created using sol-gel processing of silicates without calcination or sintering. Silica is an advantageous material for membranes because of its excellent chemical stability and extremely low cost. Furthermore, the gelation process can be utilized to tune the pore size for effective size exclusion of electrolytes (*i.e.,* good ion selectivity). In this study, small angle x-ray scattering (SAXS) was utilized to characterize the membrane pore structure. Model fitting can be utilized to extract the porosity as well as pore shape, size and size distribution. By varying the gelation conditions, average pore radii were found to be in the range of 0.4 – 2 nm. This pore structure is nearly ideal for vanadium-based RFBs given that pores must be able to effectively transport protons (radius of 0.25 nm) but not vanadium ions (radius of 0.5 nm). Nanostructure was shown to be correlated to key membrane performance attributes such as area specific resistance and optimized to be as low as industry standard perfluoro sulfonic acid (PFSA) membranes that are only 50 μm thick.

Nanoporous silica membranes were also found to have improved mechanical properties when compared to PFSA membranes. In this work, we found nearly 30x less swelling but 20% higher water sorption. These improved mechanical and handling features are critical if RFBs are to achieve fully-automated stack assembly. Furthermore, 18-month accelerated chemical oxidative stability tests show no degradation for silica-based membranes which is important for all-vanadium RFB lifetime. Finally, we demonstrate that it is possible to reach the same energy efficiency as PFSA membranes in an all-vanadium RFB system under common industry cycling conditions. These promising results highlight the commercial potential for a new RFB membrane that is inexpensive, chemically robust, mechanically durable and ion selective.

576902 Fluorenone Based Anolyte for an Aqueous Organic Redox-Flow Battery

Here we present the first aqueous organic redox-flow battery based on a fluorenone anolyte. The use of polar carboxylic acid functional groups on the core fluorenone molecule enables good solubility under alkaline conditions. Three isomers of carboxyl functionalized fluorenone were screened, with the most promising, 9-fluorenone-2- carboxylic acid, achieving a maximum solubility of 0.8M in 1M KOH with a redox potential of -690 mV vs SHE. When paired with a ferro/ferricyanide redox couple, an open cell potential of 1.18 V was achieved. However, during charge/discharge cycling significant capacity loss was observed within the first few cycles. In addition, flow-battery tests showed low limiting current densities and peak power densities. Subsequent analysis revealed that the redox reversibility of the fluorenone molecules is strongly dependent on pH and potential, which can lead to undesired irreversible redox reactions.

576381 Experimental Investigation of $YMnO_3$ Perovskites with Sr a-Site and Al B-Site Doping for Solar Thermochemical Fuel Production

Two-step metal oxide redox cycles employ relatively simple chemistry and enable thermodynamically favorable reduction of H_2O or CO_2 to produce H_2 or CO, respectively, at lower temperatures than allowable by thermolysis. In a typical temperature-swing cycle, the metal-oxide is reduced at high temperatures and low oxygen partial pressure, followed by re-oxidation in the presence of H_2O or CO_2 at a lower temperature. Ceria is considered to be the current state-of-the-art metal oxide due to its rapid reaction kinetics and high entropy changes during oxygen exchange, the latter of which enables thermodynamically favorable oxidation even with small temperature swings and low oxidant concentrations. However, due to its large enthalpy change during oxygen exchange, it requires high temperatures (*e.g.* 1773 K) or very low oxygen activity (*e.g.* 10^{-5} atm) to achieve notable reduction extents, thereby lowering its total fuel yield per cycle.

Many recent efforts in the thermochemical community have focused on lowering the reduction temperature through the use of emerging materials such as perovskite-oxides, of the general form ABO_3, which reduce more easily and, thus, have a higher oxygen exchange capacity. $YMnO_3$ perovskites with Sr^{2+} doping ($Y_{1-x}Sr_xMnO_3$) have been shown to achieve higher reduction extents than both ceria and similarly doped $LaMnO_3$ perovskites ($La_{1-x}Sr_xMnO_3$), and may produce more fuel depending on the oxidation conditions. The Sr^{2+} concentration is directly related to the thermodynamic properties of the material. Reduction extents were shown to increase with increasing Sr^{2+} incorporation; however, this improvement comes with the requirement of larger temperature swings and oxidant concentrations to perform the oxidation reaction under thermodynamically favorable conditions. Additionally, doping of Al^{3+} in $La_{1-x}Sr_xMnO_3$ was shown to lead to a higher oxygen exchange capacity as a result of Mn-enrichment on the surface.

Herein, we present an investigation of $Y_{0.8}Sr_{0.2}Mn_{0.6}Al_{0.4}O_3$ (YSMA8264), $Y_{0.8}Sr_{0.2}Mn_{0.4}Al_{0.6}O_3$ (YSMA8246), and $Y_{0.9}Sr_{0.1}Mn_{0.6}Al_{0.4}O_3$ (YSMA9164). Assessments of the oxygen exchange capacities of these materials via temperature programmed reduction showed a significant improvement in reduction extents when compared to ceria. Equilibrium measurements were performed under conditions typifying the reduction and oxidation steps, respectively. Thermogravimetric analysis was employed to measure the nonstoichiometry of the materials at temperatures from 1173-1473 K and oxygen activity from 1.61×10^{-4}-3.23×10^{-2} atm. To assess the viability of low temperature oxidation, an isothermal relaxation scheme was used in a custom tubular reactor. The measurements were performed at temperatures from 973-1173 K and oxygen partial pressures from 1.23×10^{-20}-2.24×10^{-13} atm. During these experiments, the oxygen partial pressure was controlled by delivering a precise mixture of H_2O and H_2, with H_2O:H_2 ratios ranging from 1.51-72.54. Under high oxidant-to-fuel conversion conditions (*i.e.* H_2O:H_2 ratios approaching 1), relatively small changes in oxygen content compared to ceria were observed for each of the materials.

573522 Understanding the Oxidation Kinetics Mechanism of Ceria Based Materials in Solar Thermochemical Gas Splitting

Ceria is commonly doped with transition metal and rare earth, particularly Zr, Hf, Pr and Gd, to increase the product yield and or stabilize gain sizes during solar thermochemical water and carbon dioxide splitting. In these cycles the neat or doped ceria is heated via concentrated sunlight to over 1400°C, causing the material to reduce and release O_2; subsequently the material is exposed to steam or carbon dioxide to re-oxidize the ceria and generate the desirable H_2 or CO. The use of dopants including Hf and Zr have proven to be useful in increasing the overall cycle efficiency while Pr and Gd help to stabilize the phases. Here we use density functional theory, and kinetic Monte Carlo simulations to explore the effects of these dopants on the kinetics of re-oxidation. We find that the reactions are not diffusion limited but rather surface limited, being either sub-surface/surface O migration, or H2 formation limited, depending on the dopant. These results correspond with existing experimental results and provide an explanation for the surface limited reaction.

571837 Functional Ceramics for Solar Thermochemical Storage, Hydrogen and Liquid Fuel Production

Functional ceramics for solar thermochemical storage, hydrogen and liquid fuel production

The intermittent nature of renewable sources is one of the biggest challenges to overcome in their widespread utilization. In case of solar energy a promising method is to store it in the form of chemical energy in compounds such as hydrogen or hydrocarbons. These products are useful as storage medium, as energy carrier, as a "solar fuel" or as an intermediate for chemical commodities. Among the fuels, hydrogen is of specific interest, since the only product of hydrogen combustion is water, which in turn is the feed to produce hydrogen with solar energy.

Thermochemical cycles utilize a combination of chemical reactions driven by heat to thermochemically decompose water; therefore they are a viable option to produce hydrogen at acceptable temperature levels. The two-step cycles can achieve high efficiencies, and the materials are noncorrosive. The high temperature, therefore thermodynamically limiting step is the reduction of the oxidized form of the redox material, while the water splitting step takes place at lower temperatures as shown in Fig. 1. The most prominent examples are redox cycles applying metal/metal oxide pairs (such as ZnO/Zn), multi-valent metal oxides (e.g. Fe_3O_4/FeO, CeO_2/Ce_2O_3), or a metal oxide with oxygen non-stoichiometry (perovskites), which are capable to reversibly incorporate oxygen in their lattice.

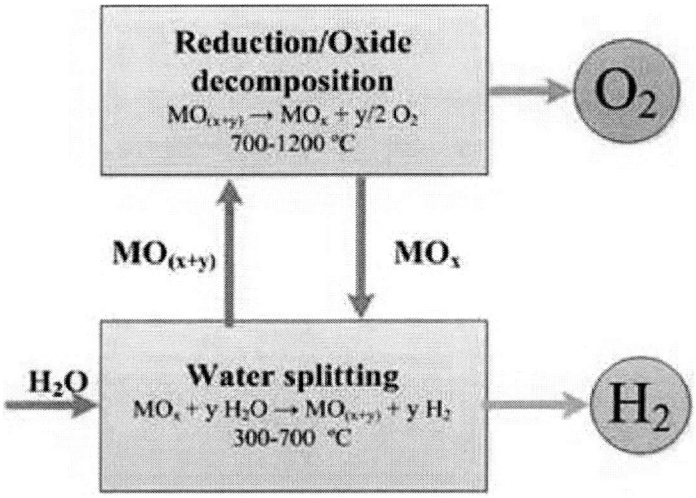

Fig. 1. Scheme of metal/metal oxide cycles for thermochemical hydrogen production.

Thermodynamic studies showed that ceria-based as well as perovskite materials are viable candidates as redox materials for thermochemical water splitting cycles. Perovskites are of special interest since they undergo a partial reduction without phase change; therefore they are more durable for long-time applications. The AMO_3 perovskite structure can be achieved by using different A and M ions, which their thermodynamic properties tuneable. In the scope of this work a library of redox materials with different compositions was established via large-scale material screening. The most promising compositions were subjected to equilibrium reduction and oxidation experiments under varying conditions to study the thermodynamic properties. Further experiments were conducted to achieve a basic understanding of interrelations between atomic structure, transport mechanisms, microstructure, reactivity and life time of the materials. The chosen functional ceramics were further developed to define and create suitable designs for solar receivers, chemical reactors and process schemes.

In parallel to material development optimizing the performance of the equipment needed to run such a process in a solar tower facility is essential. The use of particulate redox materials has been proposed to decouple the reduction reaction from the oxidation reaction, to decouple the absorption of solar radiation from the reduction reaction by using inert heat transfer particles and to allow a counter current heat exchanger design by using a combination of both particles. This concept can be implemented into any particle process and offers potentially high heat recovery rates and thus it can improve the overall system efficiencies. A detailed model to calculate the performance of the concept in consideration of temperature dependent material data and several other influencing factors is developed and validated by experiments.

Summary

The present work aims to give an overview on recent developments and the state-of-the art of high temperature fuels solar processes, in particular hydrogen production via solar thermochemical cycles. It will have a look on the most important performance parameters involved and will give an outlook on further potential and necessary developments.

572869 A Comparison of Solid-Gas Thermochemical Energy Storage Reactions and Process Configurations for CSP Applications

Concentrating solar power (CSP) integrated with thermal energy storage is considered to be a promising option to deliver cost effective and dispatchable renewable power. Thermochemical energy storage (TCES), which is the reversible conversion of solar-thermal energy to chemical energy, allows high operating temperatures, high storage density and low heat loss over long periods. Many fluid-phase and solid-gas reactions have been studied as candidates for TCES.(Pardo et al., 2014) Recently, solid-gas TCES reactions have been receiving more attention because they require less gas storage, have higher reaction temperature and easier product separation.

Today, TCES is still at an early stage of development. While considerable work has been done in reaction selection and reactor design, there are limited system-level studies for the design of integrated processes with TCES systems embedded in CSP plants. As a result, it remains unclear what is the best process configuration, reactor design, and chemistry for CSP systems with TCES. Accordingly, the goal of this work is to (1) explore process configurations for solid-gas TCES (2) evaluate the performance of different TCES strategies via optimization models, and (3) determine how the characteristics of a reversible reaction affect the selection of a TCES process configuration.

The key units of a TCES process are two reactors: R1, where the solar heat drives an endothermic reaction ( ) and R2, where the reverse exothermic reaction takes place and the reaction heat is used for power generation. To systematically develop TCES processes, we identify four key process characteristics with two options available for each: (1) R1 heat receiving (directly from solar irradiance/ indirectly from heat transfer fluid (HTF)), (2) heat transfer between R2 and power cycle (working fluid direct/ indirect contact with reactants), (3) reactor operation mode (continuous/ batch), and (4) presence of gas storage (closed-loop/ open-loop). The selection of an option for each characteristic leads to a configuration. After a preliminary screening on the possible combinations, we select the six most promising configurations for this analysis.

A TCES strategy is defined by the combination of a reversible reaction and a process configuration. We select three reactions, each representing one reaction group: (Carbonate group), (hydroxide group), and (redox group). Carbonate and hydroxide reactions can only be paired with closed-loop configurations, while redox reactions can be combined with all six configurations. Therefore, the three reactions and six configurations selected in this work lead to 12 TCES strategies in total.

We develop optimization-based process synthesis models to investigate the energetic and economic performance of these 12 strategies (Peng et al., 2017). We build models for different TCES units. Special emphasis is placed on the modeling of fixed-bed reactors, which operate in a cyclic batch mode and their first-principal model involves partial differential equations (PDEs) in time and space. Solving this first-principal model, especially for optimization, is computationally expensive, so we build surrogate models with simple algebraic equations to replace the complex first-principal models. Specifically, we first identify key input and output (e.g. heat storage and reaction conversion) variables; then generate data points using expensive numerical simulations implemented in COMSOL (which are carried out offline); generate surrogate models via data fitting; and, finally, integrate the surrogate models into the CSP-TCES process synthesis optimization model.

To design the plant under seasonal and daily variability in solar irradiance, we formulate the optimization problem as a two-stage stochastic programming model with day and night modes nested inside each scenario (Peng et al., 2018). By solving the model, we obtain the first stage design decisions (e.g. solar field area and all equipment sizes) and second stage operational decisions (e.g. stream flow rates, unit operating conditions) that yield the minimum plant levelized cost of electricity (LCOE). Other performance metrics such as energy efficiency and storage density of the optimal design are also calculated.

The analyses allow us to identify cost-efficient TCES systems. Among the 12 strategies, the hydroxide reaction + configuration with direct irradiated R1 and continuous reactor shows the best economic performance, achieving a 20% LCOE reduction over the commercial two-tank molten salt CSP plants. From this analysis, several general design rules and insights are also generated to serve as guidelines for the future development of TCES processes.

Reference

Pardo, P., Deydier, A., Anxionnaz-Minvielle, Z., Rougé, S., Cabassud, M., Cognet, P., 2014. A review on high temperature thermochemical heat energy storage. Renew. Sustain. Energy Rev. 32, 591–610. https://doi.org/10.1016/j.rser.2013.12.014

Peng, X., Root, T.W., Maravelias, C.T., 2018. Optimization-based Process Synthesis under Seasonal and Daily Variability: Application to Concentrating Solar Power. AIChE J. 0, 1–16. https://doi.org/10.1002/aic.16458

Peng, X., Root, T.W., Maravelias, C.T., 2017. Storing Solar Energy with Chemistry: The Role of Thermochemical Storage in Concentrating Solar Power. Green Chem. 2427–2438. https://doi.org/10.1039/C7GC00023E

564517 Flexible Heat Integration to Enhance Solar Energy Utilization in a Hybrid CSP Plant

Concentrated solar power (CSP) represents a potential means to harness and dispatch solar energy on a large, centralized level. However, like other renewable energy sources, CSP must deal with issues of transiency and flexible operation if it is to be competitive in the modern power grid. Due to its ability to hybridize with fossil fuels through equipment commonality and integrate thermal energy storage (TES), these issues can be mitigated. Further penetration can be achieved by employing holistic automation of CSP systems to leverage TES and hybrid operation. This work proposes a novel operation scheme, referred to here as flexible heat integration (FHI), to enhance solar energy utilization in a solar central receiver—Rankine cycle (SCR-RC) power plant hybridized with natural gas. Typically, in an SRC-RC, concentrated solar energy heats a molten salt and dispatches the salt to a series of heat exchangers in the following order: the steam superheater, steam generator, and feedwater preheater. Under variable conditions (low- and medium-grade solar activity), maintaining the high salt temperatures needed for steam superheating can in result dips in collector thermal efficiency and by extension, diminished solar energy utilization. Instead, FHI explores dispatching solar energy independently to the thermal sinks based upon solar activity rather than in sequence while having hybrid operation via natural gas firing make up for the bypassed energy sinks. For example, during low solar activity, solar energy can be more efficiently collected and stored at a lower temperature and then dispatched to the correspondingly low-temperature feedwater solar preheater while gas firing generates and superheats steam. Likewise, during medium-grade solar activity, a solar salt heated to a medium-level temperature can dispatched to the steam generator and subsequently the feedwater preheater while gas superheats steam. Under such a design, additional TES tanks are needed beyond the conventional two-tank system seen in current SCR-RC power plants. This study is conducted to detail the fundamentals of FHI and how it can be employed to further improve solar energy utilization within a hybrid SCR-RC system thereby increasing CSP penetration and reducing greenhouse emissions relative to the standard SCR-RC power plant.

579318 Gasification of Organic Solid Waste Under High-Flux Solar Irradiation: Thermodynamics, Kinetics, and Fixed-Bed Reactor Evaluation

Conventional autothermal gasification processes are driven by heat derived from combustion of a portion of feedstock. Combustion generates byproducts, which contaminate the product synthesis gas (syngas) – a mixture mainly consisting of H_2 and CO. The advantages of solar-driven gasification as a means of upgrading carbonaceous feedstocks to syngas were demonstrated in prior work in the field. Here, we present a study of solar thermal dry gasification of organic waste including thermodynamics, chemical kinetics and design, fabrication and evaluation of a 2 kW_{th} fixed-bed CO_2 gasifier. Thermodynamic analyses were conducted to explore theoretical efficiency and composition limits of the process. Chemical kinetics of CO_2 gasification were investigated by employing thermogravimetric analysis with selected types of carbonaceous feedstocks including waste wood and sewage sludge. The high-flux solar irradiation tests were performed with a windowless solar gasifier with two concentric silicon carbine cavities: the inner cavity serving as a solar receiver to absorb high-flux solar irradiation and the outer cavity serving as a reactor chamber for gasification. The gasifier tests were performed under simulated sunlight provided by a new multi-source high-flux solar simulator developed at National University of Singapore. The solar flux distribution in the reactor aperture plane was measured by a water-cooled thermogage heat flux sensor and a CCD camera. The effects of CO_2/C ratio and input solar power to the reactor on the carbon conversion, cold gas efficiency, and solar-to-fuel efficiency were experimentally explored.

570223 Analysis of the Effect of Solar Pyrolysis Parameters on the Physicochemical Properties of Biochar

Carbon materials can be produced by different methods. One of them is solar pyrolysis, which is attaining renewed interest due to simplicity to transform biomass with environmental advantages. However, pyrolysis operation parameters are decisive factors that determine the morphological and chemical properties of carbon.

This work aims to study the influence of temperature and heating rate of solar pyrolysis in the properties of carbon. Pyrolysis experiments were carried out in the 25 kW solar furnace of IER-UNAM. *Agave angustifolia* biomass was processed in a glass spherical solar reactor. Temperatures between 500 and 900°C were employed, as well as heating rates between 4 and 30°C/min. The char produced was analyzed with several techniques, which include elemental analysis (CHONS), surface area by BET and DFT models and capacitance by cyclic voltammetry.

The results of the experiments suggest that solar pyrolysis temperature and heating rate scarcely impact the carbon composition, nevertheless the structure, surface area and electrochemical response are highly affected by both. The control of the operation conditions allows to produce carbon materials that exhibit promising properties to be used as catalyst supports for polymer electrolyte membrane fuel cells and electrolyzer cells applications. Consequently, the use of concentrated solar energy represents an attractive method to produce these valuable materials.

567177 Numerical Design and Experimental Evaluation of a Direct Solar Reactor for the Catalytic Dry Reforming of Methane

Methane dry reforming is a process to produce "syngas" (H2 and CO) which is a major precursor for many chemicals and ultra-clean fuels. It is also a CO2 assisted process that leads to the conversion of CO2 to higher value products and an overall reduction of CO2 emissions from the chemical industry. For this reason, there is a good number of studies and reviews towards improving the process efficiency, designing new catalysts, and utilizing solar energy. This study is part of a bigger effort that explores novel catalysts and processes for CO2 conversion, with the goal of achieving processes in which CO2 input is greater than CO2 output at any stage, with the ultimate objective of significantly lowering Qatar's CO2 footprint. This study focuses on the utilization of solar energy to activate the Atomic Layer Deposition (ALD) based catalyst and drive the dry reforming reaction directly. First, we present the numerical design of a lab scale solar tubular reactor and second, the experimental evaluation of the reactor and dry-reforming reaction conversion efficiencies.

The thermal and optical performance of a solar tubular reactor/receiver depends upon the geometry of the receiver, optical and thermophysical properties of a receiver's material and reaction conditions. Initially, out 7kW high flux solar simulator (HFSS) is modeled using a commercial software, TracePro®, to obtain the intensity and distribution of the incident radiation flux. The tubular receiver has an internal diameter of 20 mm and an "active" length of 20 mm. To achieve the maximum possible optical efficiency and homogenous heating of the receiver a novel cavity is designed which reflects and focuses the incident flux on the tubular receiver. Multiple scenarios were investigated and optimum results are presented here.

Following the solar reactor manufacturing, an experimental setup was fabricated to study the performance of the proposed catalysts under simulated on-sun conditions. Two different catalysts have been used in a fixed bed configuration (0.2 g catalyst premixed with 2 g of inert quartz), a commercial and an ALD-based Ni/Al_2O_3. The gas mixture (CH_4, CO_2, Ar, He with 5%: 5%: 1%: balance) was fed into the reactor at weight hourly space velocities (WHSV) in the range of flow 2,000 ml/min/g to 10,000 ml/min/g. The catalytic bed was directly irradiated with a high flux solar simulator using 1-5 Xe arc lamp (6 kW each) producing three different temperature levels (550 °C, 650 °C and 850 °C). Herein we report the methane reforming conversion efficiencies and catalysts' activities. Also, the morphological, chemical, and stability changes of the catalytic particles have been studied and reported here.

571234 Economic and Dynamic Performance Solar-Driven Thermochemical Production of Hydrogen Using Ceria

Economic and dynamic performance solar-driven thermochemical production of hydrogen using ceria

Alicia Bayon and Alberto de la Calle

CSIRO Energy, PO Box P.O. Box 330, Newcastle, NSW 2300, Australia

In this work, we explore the dynamic performance of a 1 MW_{th} hydrogen production plant based on ceria thermochemical cycle. This work extended from our previous dynamic modelling recently published (de la Calle and Bayon 2019) by incorporating an economic analysis of the production of hydrogen. In this presentation, we aim to explain the main critical characteristics of the plant operation that affects the overall solar-to-fuel performance: heat recovery strategy, location, heliostat field operation and thermal inertial of chemicals and components.

The elements of the plant are shown in Figure 1. The main models are: a data source for encapsulating the weather data extraction, a heliostat field layout optimized using SolarPILOT^{TM}, a rotatory cavity receiver, two tanks for storing Ceria particles, two particle conveyors that allows Ceria to move between the reactors (labelled CeO_2 pump 1 and 2), gas valves to control the pressure in the reactors, heat exchangers and two high temperature membranes for separating the product gases. The operation of the plant begins by transporting solid particles of CeO_2 from a tank (CeO_2 tank 1) to the receiver (Figure 1) when the solar energy is available. CeO_2 is reduced using a flow of N_2 to enable a low O_2 partial pressure atmosphere. In our design, the reduced Ceria (CeO_{2-d}) is stored (in the CeO_2 tank 2) at a high temperature in order to allow a continuous flow of hydrogen. The gases coming into the receiver and oxidizer are pre-heated using the gas outlet and with heat exchangers (N_2 HEX and H_2O HEX). Heat recovery is also implemented over the ceria particles moving from the oxidizer to the receiver (CeO_2 HEX). In addition, the pressure in the reactors is controlled with purge valves. For accomplishing this study, a library of typical meteorological year was developed for ten location selected.

The plant is located in Learmonth (Western Australia), selected from our previous work as the most efficient location close to a maritime port (Bayon and de la Calle 2019) showing that a maximum production of hydrogen of 17.46 tons a year. From this plant design, we have performed a further techno-economic analysis showing that the technology could produce hydrogen at a selling price of \$18 GJ^{-1} if the total Capex is around 9 million \$ and CO_2 reduction incentives are considered. We also compared our results with previous techno-economic analysis with other forms of hydrogen like electrolysis and conventional reforming. This analysis will also provide the selling price of the hydrogen and maximum Capex and Opex to be achievable for deployment this technology with a minimum net present value of \$1 million over 30 years of operation together with a maximization of the internal rate of return.

This presentation will show how the incorporation of our economic analysis into the dynamic modelling is critical to optimize the process design conditions (e.g. particle storage inventory, maximum operational temperature, size of tanks and receivers, among others). We believe this novel approach can drive to the full economic optimization of a solar thermochemical hydrogen production plant design that can be applicable to other materials for a full comparison.

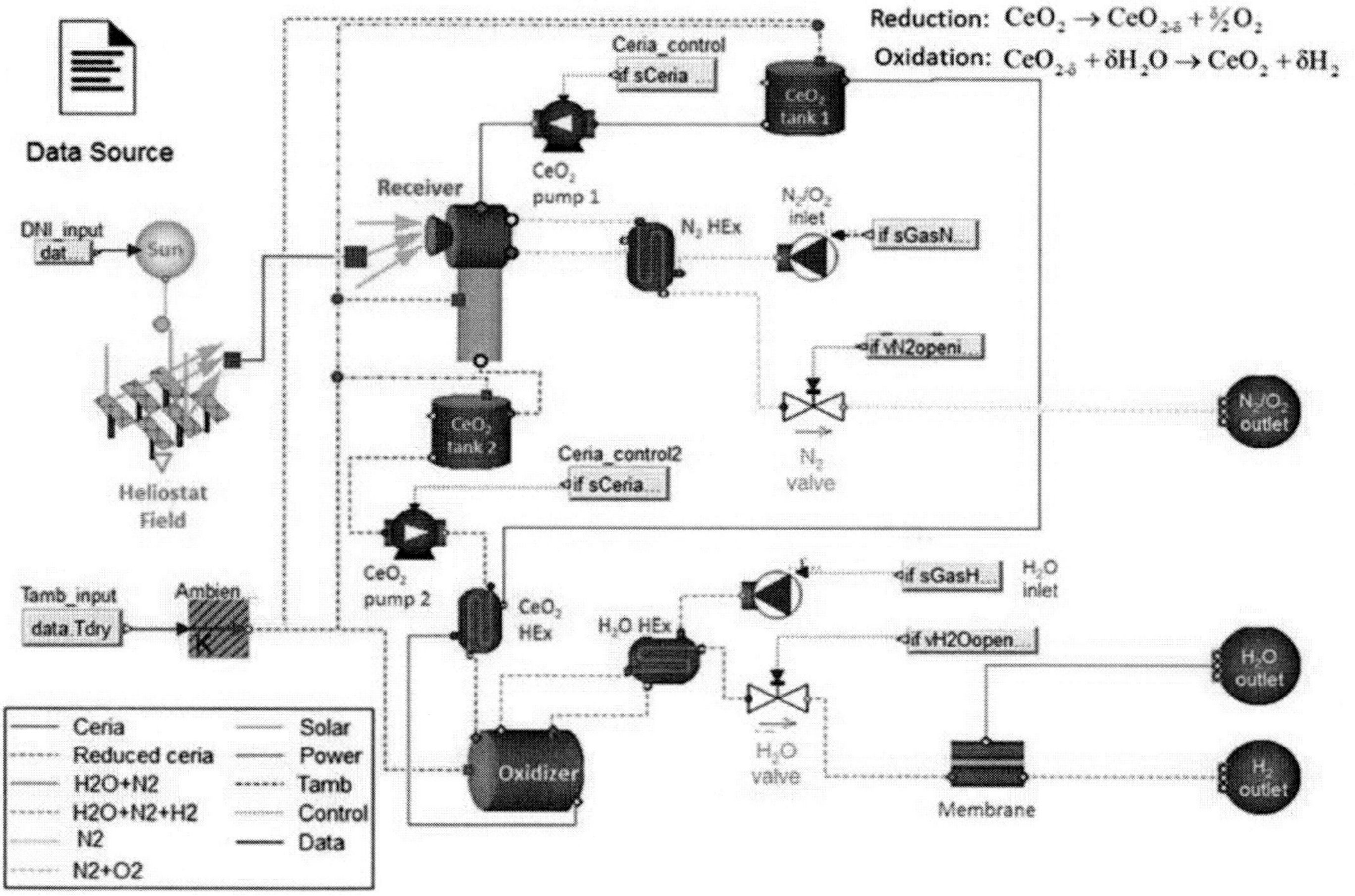

Figure 1: Modelica diagram of the solar-driven thermochemical production plant based on ceria.

Bayon, A., Calle, A. de la, 2019. Effect of Plant Location on the Annual Performance of a Hydrogen Production Plant based on CeO2 Thermochemical Cycle. AIP Conf. Proc. - SolarPaces2018 Accepted.

de la Calle, A., Bayon, A., 2019. Annual performance of a thermochemical solar syngas production plant based on non-stoichiometric CeO2. Int. J. Hydrogen Energy 44, 1409–1424. https://doi.org/10.1016/J.IJHYDENE.2018.11.076

564021 Dish-STARS™ Solar Thermochemical Production of Hydrogen

Dish-STARS™ Solar Thermochemical Production of Hydrogen

Richard F. Zheng [1], Robert S. Wegeng [2], Paul H. Humble, Johnny Saavedra-Lopez [1], Timothy G. Veldman [1]

[1] Pacific Northwest National Laboratory, Richland, WA, United States

[2] STARS Technology Corporation, Richland, WA, United States

The present paper describes the recent advances in solar thermochemical processing and reactor engineering to achieve greater than 70% solar-to-chemical energy efficiency, based on PNNL's Dish-STARS™ technology. In a Dish-STARS™ module, a Solar Thermochemical Advanced Reaction System (STARS) is coupled to a parabolic dish solar concentrator to convert solar energy into storable/useful chemical energy. The dish concentrator has been previously developed for electrical power generation. The STARS unit is a compact, process-intensive chemical reaction system based on micro- and meso-channel process technology (MMPT). Dish-STARS™ steam methane reforming has been demonstrated on-sun to produce syngas at over 70% solar-to-chemical energy efficiency, while analyses indicating a potential for values greater than 80%.

The combined Dish-STARS™ system provides a solar augment to a methane stream such as natural gas, landfill gas and biogas, increasing its chemical energy content by 20-30% while decreasing its carbon intensity. If the syngas product is further reacted to produce valuable chemical products, such as hydrogen, the reduction in carbon emissions is retained and reduced carbon intensities can be attained for the chemical products. In a co-production mode, low-carbon hydrogen or electricity plus various hydrocarbons (for example, methanol, olefins or plastics) can be produced. Currently, the Dish-STARS™ technology is under commercial development for the efficient use of concentrated solar energy in near-term applications for electrical power generation and the production of chemicals, including hydrogen (see Figure 1).

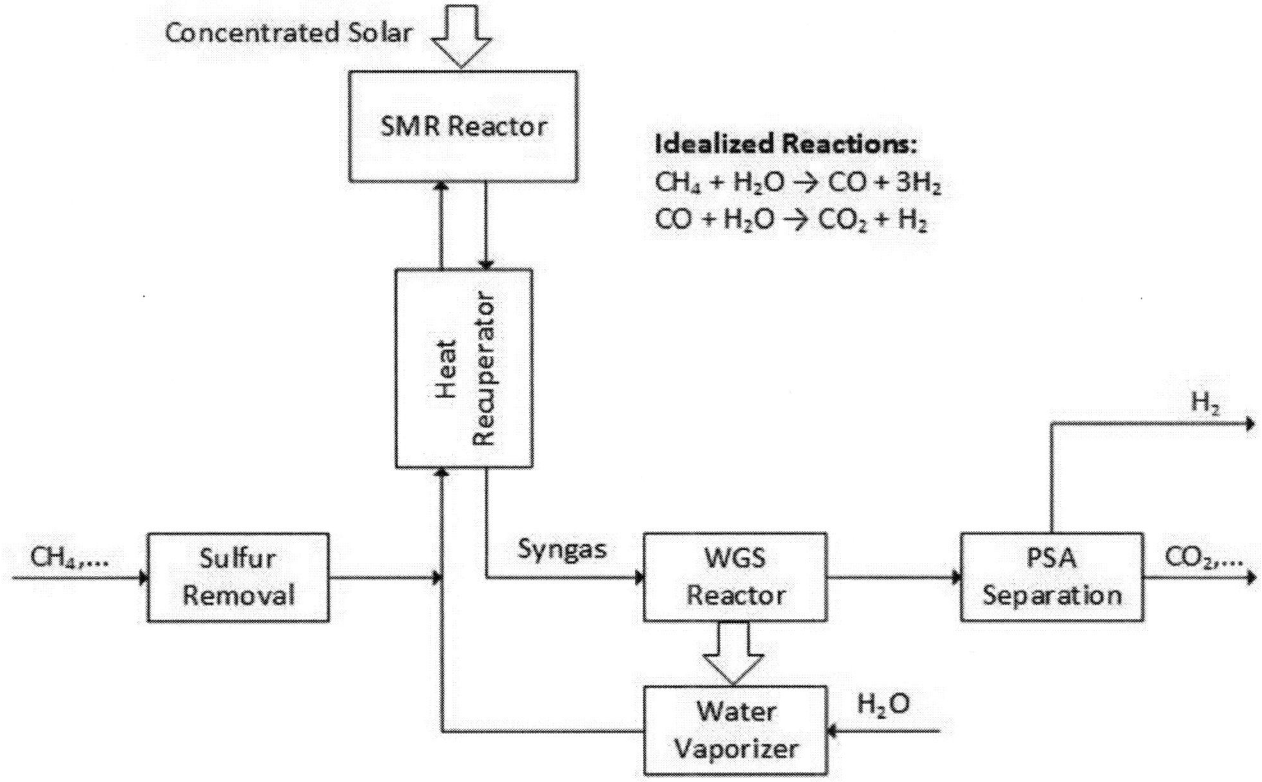

Figure 1. Level-1 block diagram of general Dish-STARS™ hydrogen production concept.

Over the course of a previous SunShot project, the STARS concept has been advanced to a technology readiness level (TRL) of 6. Reactor design was focused on improving exergetic efficiency of the SMR reactor and reducing thermomechanical stress through CFD modeling and on-sun experimental validation. A high degree of heat integration was achieved by using a number of carefully designed MMPT heat exchangers to recuperate high grade heat from the hot syngas stream. The new reactor and heat exchanger assembly was tested on-sun under high flux conditions (see Figure 2) and showed improved efficiency over previous version [1].

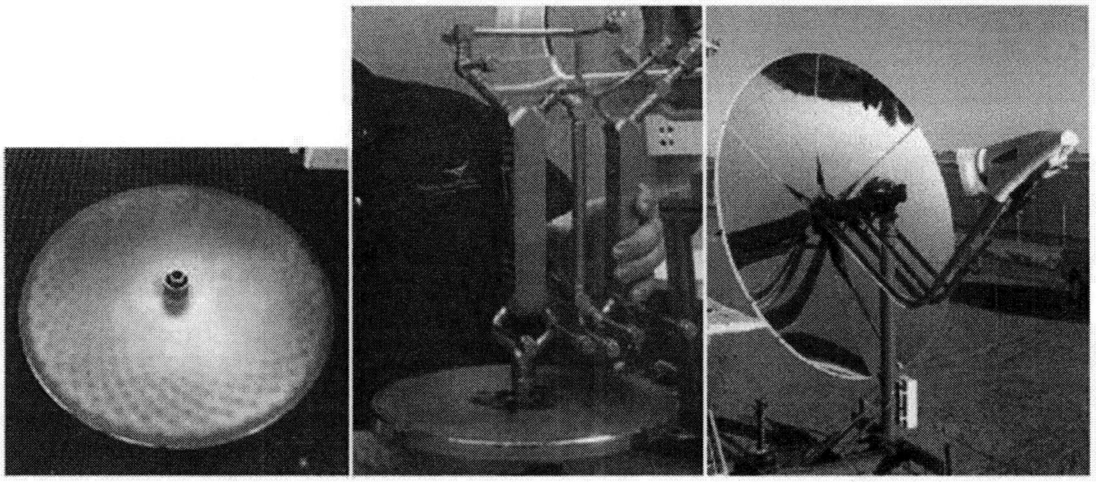

Figure 1. TRL-6 STARS reactor (left), reactor and heat exchanger assembly (middle), and the Dish-STARS™ module operated on-sun (right).

In this paper, details of the STARS reactor design to promote thermal spreading and reduce thermomechanical stress will be discussed. On-sun steam methane reforming performance data and system exergy analysis results will also be presented. Preliminary results from a manufacturing investigation will be included.

References

[1] Zheng, R.F. et al., *Energy Procedia* **69** (2015) 1192.

572469 Incorporation of a Phase Change Material into Fibers for Thermal Modulation and Energy Storage

Effective thermal modulation and storage is one of the most important issues for achieving energy-efficiency across all sectors. Phase change materials (PCM) can act as heat reservoirs by effectively storing or releasing massive amounts of heat during the phase change process (typically a solid-liquid transition). PCMs have been developed with a wide range of operating temperatures and integrated into various platforms such as building materials, gas sorbents/separators, and consumer products, among others. In this work, we fabricate polymer fibers that possess high loadings of PCMs to provide effective thermal modulation for several applications. Polymer fibers offer distinct benefits over other contenders since they can be solution processed and produced at enormous scales. We incorporate PCMs into various polymer dopes (e.g. cellulose acetate, polyethersulfone, etc), and PCM-polymer fibers are then spun via conventional solution-spinning processes. The thermal response behaviors of PCM-polymer fibers were analyzed by using differential scanning calorimetry (DSC), and it was confirmed that there was no damage to PCMs during the fiber spinning and no degradation of PCM after several freeze/melt cycles. The actual loading amount of PCMs in fibers can be obtained up to 80 wt%, depending somewhat on the dimension of fibers. Dynamic mechanical analysis (DMA) reveals that there is a trade-off between mechanical stability of PCM-polymer fiber and loading amount of PCMs, thus requiring optimization of the PCM loading to meet application-specific mechanical stability targets. We expect our engineered PCM-polymer fibers can be applied to a smart thermal energy storage material that enables effective heat management.

577684 Latent Heat Trapping in Phase Switching Liquids

Latent Heat Trapping in Phase Switching Liquids

Preventing ice from forming on functional surfaces is a major engineering concern. Industrial strategies to mitigate such hazards include active mechanical, chemical, and electro-thermal deicing techniques that are energy and cost intensive. In the current study, we show that the use of certain phase change materials, which are in liquid state under ambient conditions and have melting point higher than the freezing point of water, referred herein as phase-switching liquids (PSLs), can delay condensation–frosting lasting up to 300 longer in bulk state compared to conventional surfaces under identical environmental conditions. The freezing delay is primarily a consequence of the release of trapped latent heat due to condensation, but is also affected by the solidified PSL surface morphology and its miscibility in water. We show that regardless of surface chemistry, PSL coated surfaces exhibit low droplet adhesion when operated below the corresponding melting point of the solidified PSLs, engendering ice and frost repellency even on hydrophilic substrates. Finally, we discuss how solidified PSL surfaces can display varying degrees of optical transparency, can repel a variety of liquids, and self-heal upon physical damage.

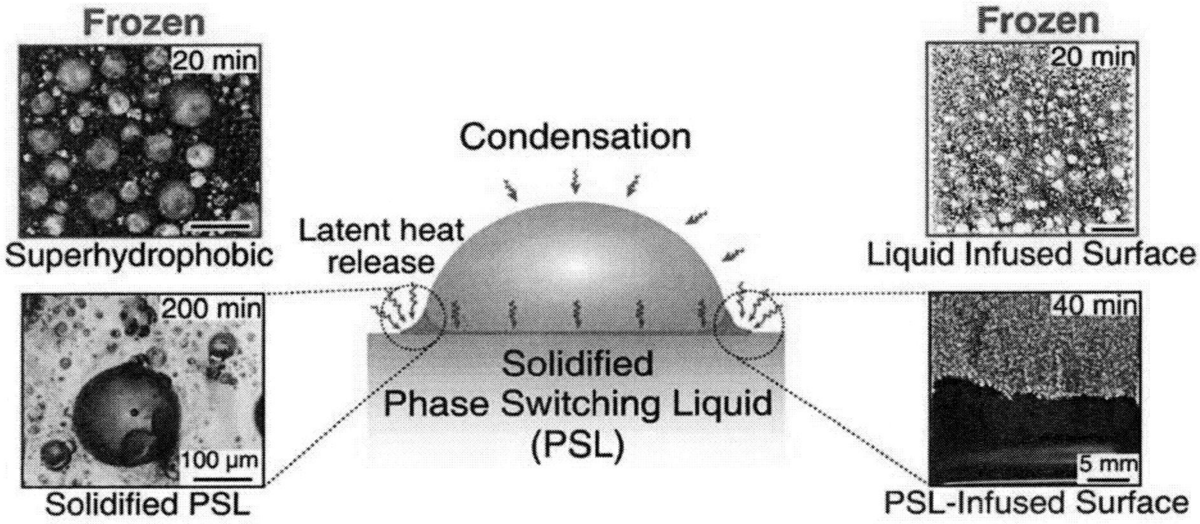

Figure 1. Exceptional anti-icing performance by latent heat trapping in Phase Switching Liquids

580389 High Pressure High Temperature Thermal Energy Storage Systemfor Concentrated Solar Power (CSP) Plants

A High Pressure/High Temperature (HPHT) Thermal Energy Storage (TES) system based on inexpensive pebbles as the heat transport medium will be discussed after reviewing low pressure TES systems under development and undergoing field tests. The particle-bed TES design developed at City College of New York in 2011 used 8 mm diameter alumina particles as the heat storage medium contained in a 10' long, 8" diameter stainless-steel pipe and heated up to 600 ^{o}C at 30 bar. The thermal storage capacity was, however, limited due to the small internal volume of the stainless-steel pipe that could be adequate for small-scale use. To scale up this design to become practical for large scale Concentrated Solar Power plants that can produce high pressure air or superheated steam from a solar receiver, a large-scale TES system needs to be designed without raising the capital cost. For CSP plants equipped with Direct Steam Generation (DSG) solar receivers, for example, a 5 MWth CSP plant built and operated in Turkey, the latent and sensible heats of high pressure, superheated steam or sensible heat of a high pressure, high temperature gas needs to be stored in the pebble-bed TES system at pressures above 60 bar and temperatures above 600 ^{o}C. Considerations of design features needed to reduce the capital cost but can meet the TES requirements will be discussed.

572212 Flame-Retardancy and Thermal Properties of a Novel Phosphorus Modified PCM for Thermal Energy Storage

The organic phase change materials are flammable, which limit their applications in heat storage and conversion. In the presented study, a phosphorous-grafted hexadecanol was blended with PEPA to prepare flame-retardant phase change materials (FRPCM). The thermal properties of FRPCM were evaluated by means of differential scanning calorimetry (DSC), thermogravimetric analyses (TGA), cone calorimeter and scanning electron microscope (SEM) tests. The TGA revealed that grafted phosphorous group helps to promote char formation and improve decomposition temperature. The DSC results showed that FRPCM maintained a promising heat storage capacity (186.9 J/g) and thermal stability after grafting and blending. The cone calorimeter test demonstrated that the peak heat release rate and total heat release were further decreased by more than 50%, owing to the synergistic effect from phosphorous group and PEPA. The SEM results showed that porous structure was formed during combustion. The flame retardance mechanism was also discussed.

Experimental Study on Thermal Performance of a Latent Heat Thermal Storage Cell Unit Utilizing Finned Metal Foam

*Yongping Huang [1], Yingying Chen [2], Liangyu Wu [2, *]*

[1] School of Energy and Environment, Southeast University, Nanjing 210096, P. R. China

[2] College of Electrical, Energy and Power Engineering, Yangzhou University, Yangzhou 225127, P. R. China

**Corresponding author: Email: lywu@yzu.edu.cn*

Abstract

The present paper experimentally studies the melting performance of finned metal foam (FMF) latent heat thermal storage (LHTS) cell units. The dynamic thermal performance of a FMF LHTS cell unit is analyzed and compared with the traditional LHTS cell unit. The effects of porosity and pore density on the melting performance of FMF LHTS cell units are analyzed. The results show that the best melting performance is achieved for the FMF LHTS cell unit since the thermal conduction enhancement exceeds the heat transfer loss arising from the suppression of natural convection. The decrease in porosity leads to energy charging enhancement and thermal storage capacity reduction. Moreover, the increase of pore density results in the melting performance enhancement without loss of the thermal storage capacity. In general, for maximizing melting performance with the least loss of thermal energy storage capacity, the porosity and pore density are recommend to be 0.9 and 40 PPI, respectively.

Introduction

The solar energy is especial renewable energy with advantages of rich sources, clean and pollution-free, which is regarded as an ideal candidate of fossil energy for solving the increasingly serious environmental and energy issues. However, the inherent instability and intermittency of solar energy restrict large-scale applications in engineering [1]. To overcome this intrinsic deficiency, the development of energy storage technology is a key to increasing the share of solar energy in primary energy consumption.

It has been documented that the latent heat thermal storage (LHTS) is a better choice for energy storage in the field of concentrated solar power due to the advantages of isothermal and reversible nature of the storage process and high energy storage density [2]. However, the intrinsic poor thermal conductivities of phase change materials (PCMs) restrict the thermal effectiveness of LHTS units. To overcome this inherent problem, the addition of materials with high thermal conductivity, including metal foam [3, 4], fins [5, 6], graphite and fine metal particles [7], are widely utilized. Among these thermal conductivity enhancers (TCEs), the metal foam has been verified as an attractive candidate due to its simple techniques and low cost. However, the research on the melting behaviors in LHTS units utilizing finned metal foam (FMF) is few in the available literature. Therefore, it is of great significance to investigate the energy charging process of FMF LHTS units, especially the role of pore structure in melting performance.

To improve the energy charging performance, an innovative FMF LHTS cell unit is designed in this paper. A visual experiment is performed to comprehensively study the melting performance of a FMF LHTS cell unit. The dynamic temperature response of a FMF LHTS cell unit is analyzed and compared with the corresponding LHTS cell units. Additionally, the effects of porosity and pore density on the thermal performance of a FMF LHTS cell unit are examined and discussed for maximizing the energy charging performance.

Experimental descriptions

The experiment setup is illustrated in Figure 1. The experiment system includes FMF LHTS cell unit, power supply system and real-time data acquisition system. As shown in the figure, the FMF LHTS cell unit is composed of a finned heat sink, a film heater, a heat-protecting glass, and copper foams. The aluminum 6061 is chosen as the solid material and lauric acid is selected as the PCM in this paper. The dimension of the FMF LHTS cell unit is 120 mm × 120 mm × 35 mm, the fin thickness is 2 mm and the fin number is 6. The real-time data acquisition system consists of a PC, a data acquisition instrument and 5 T-type thermocouples (accuracy: ±0.1℃). To study the thermal performance of FMF LHTS cell units, the thermocouples are arranged at its bottom. Besides, the power supply system includes a DC stabilized power supply and a power meter. In this paper, a typical heat flux of 2.6 kW/m^2 is selected here to act as the heat source.

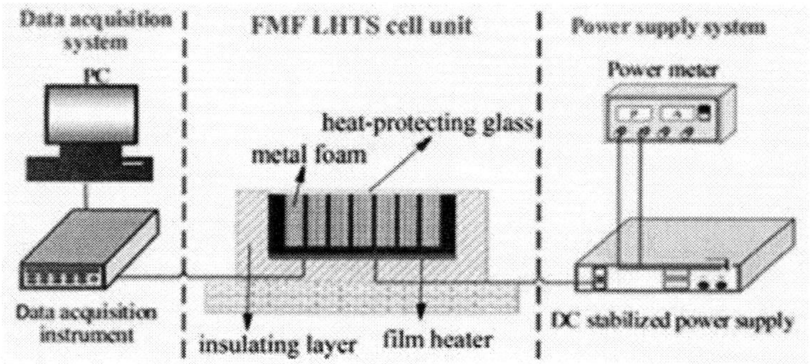

Figure 1 Experimental setup.

Results and discussions

Effect of configuration on the thermal performance

To further understand the thermal performance enhancement, the melting processes of LHTS cell units with three different configurations are studied. Figure 2 illustrates the dynamic average base temperatures of the different LHTS cell units during the melting process. As shown in the figure, the configuration of the LHTS cell unit plays an important role in melting performance. The LHTS cell unit with a cavity has the worst energy charging performance compared to the FMF LHTS cell unit and finned LHTS cell unit. Moreover, it is seen that the average bottom temperature of the FMF LHTS cell unit is lower compared to the finned FMF LHTS cell unit, which indicates that the addition of metal foam effectively enhances the melting performance of finned LHTS unit. The explanation is that the metal foam not only increases the heat-transfer area between PCM and metal matrix but also provides a fast heat flow path in the vertical and horizontal directions. In this context, the thermal conduction enhancement arising from the metal foam exceeds the heat transfer loss caused by the suppression of natural convection. Therefore, the melting performance of the FMF LHTS cell unit is the best.

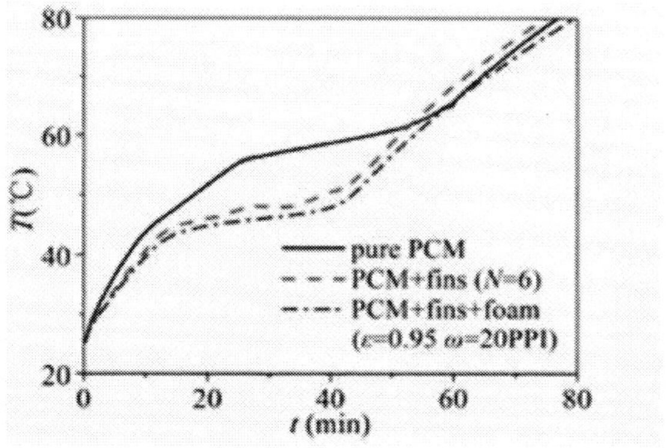

Figure 2 Effect of configurations on dynamic temperature response.

Effect of porosity on the thermal performance

To study the effect of porosity on the melting performance in FMF LHTS cell units, Figure 3 shows the average bottom temperature variations of different FMF LHTS cell units. It is seen that the increase of porosity results in the higher bottom temperature, which implies the energy charging rate is slowed down in the FMF LHTS cell units. This reason is that thermal conduction is dominant in FMF LHTS cell units due to the suppression of natural convection, and the increase in porosity weakens the effective thermal conductivity of metal foam. In addition, though the melting performance of FMF LHTS cell unit with larger porosity is worse, while the thermal energy storage capacity is enhanced. Moreover, it is interesting to find that with the further increase of porosity, the porosity plays a less important role in the energy charging performance. As the porosity increase from 0.8 to 0.9, the duration for the average base temperature of the FMF LHTS cell unit to reach a set point of 47℃is increased by 20.6%, while that is only 4.3% when the porosity further increases from 0.9 to 0.95. Therefore, the metal foam with porosity of 0.9 is recommended for larger melting performance enhancement with the least loss of thermal energy storage capacity.

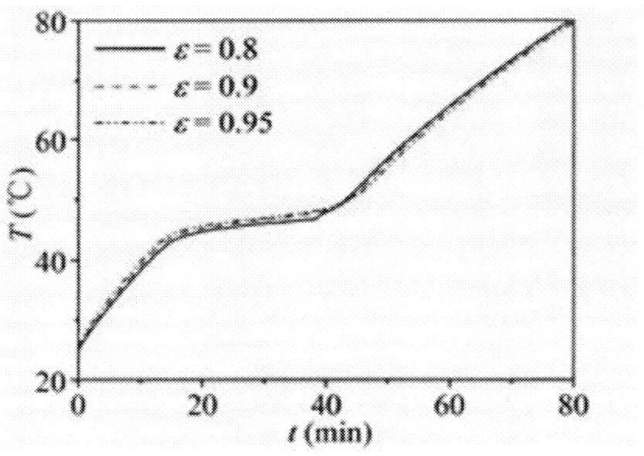

Figure 3 Effect of porosity on dynamic temperature response (ω = 20PPI).

Effect of pore density on the thermal performance

The pore density is another significant influence on the energy charging performance of FMF LHTS cell units. To study the role of pore density in the melting performance, the average bottom temperature variations of FMF LHTS cell units with different pore densities are shown in Figure 4. As shown in the figure, with the increase of pore density, the average base temperature of the FMF LHTS cell unit is lower. Moreover, since the amount of PCM is identical, the thermal energy storage capacities for FMF LHTS cell units with different pore densities are identical as well. It is noteworthy the melting performance enhancement caused by the increase of pore density is limited especially for the case when pore density is larger than 20 PPI. It is seen that the dynamic average bottom temperature of FMF LHTS cell units with w = 20 PPI agrees well with the corresponding case with w = 40 PPI. The explanation is that though the effective thermal conductivities of the different FMF LHTS cell units are the same, the increase in pore density leads to more suppression of natural convection and larger specific surface area of metal foam, and when the thermal conduction enhancement is offset by the heat transfer loss caused by the suppression of natural convection, the melting performance enhancement is restricted. Therefore, the metal foam with a pore density of 40 PPI is suggested for maximizing the energy charging performance.

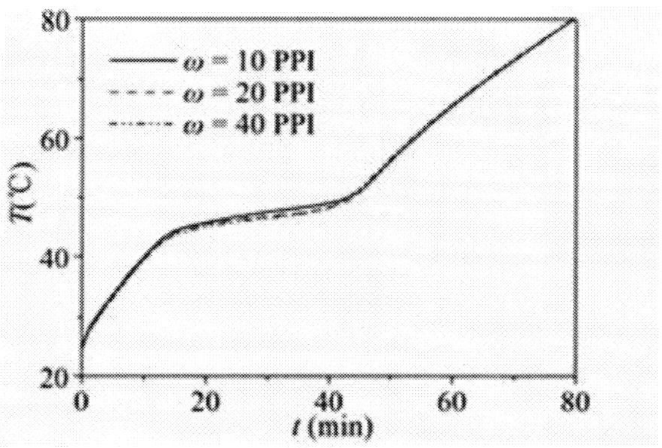

Figure 4 Effect of pore density on dynamic temperature response (ε = 0.9).

Conclusions

The effect of metal foam on the melting performance in FMF LHTS cell units is experimentally studied. The dynamic m is analyzed and compared with the traditional LHTS cell unit. The effects of porosity and pore density on the thermal performance of a FMF LHTS cell unit are analyzed and discussed. The results indicate that the configuration plays an important role in the thermal performance of LHTS cell units. The LHTS cell unit with a cavity has the worst energy charging performance, while the best thermal performance is achieved for the FMF LHTS cell unit. The porosity seriously affects the melting performance of the FMF LHTS cell unit. With the decrease of porosity, the energy charging performance is enhanced while the thermal storage capacity is reduced. Pore density also affects the melting behaviors in the LHTS cell unit. With the increase of pore density, though the thermal storage capacity is identical, the melting performance is enhanced. In general, for maximizing melting performance with the least loss of thermal energy storage capacity, the porosity and pore density are recommend to be 0.9 and 40 PPI, respectively.

References

1. A. Koca, H.F. Oztop, T. Koyun, Y. Varol, Energy and exergy analysis of a latent heat storage system with phase change material for a solar collector, Renew. Energ. 33(4) (2008) 567-574.
2. S.L. Lv, G.H. Feng, N. Zhu, D.Y. Li, Experimental study and evaluation of latent heat storage in phase change materials wallboards, Energy Build. 39(10) (2007) 1088-1091.
3. C.Y. Zhao, W. Lu, Y. Tian, Heat transfer enhancement for thermal energy storage using metal foams embedded within phase change materials (PCMs), Sol. Energy 84(8) (2010) 1402-1412.
4. C.Y. Zhao, Z.G. Wu, Heat transfer enhancement of high temperature thermal energy storage using metal foams and expanded graphite, Sol Energ Mat Sol C. 95(2) (2011) 636-643.
5. K.A.R. Ismail, C.L.F. Alves, M.S. Modesto, Numerical and experimental study on the solidification of PCM around a vertical axially finned isothermal cylinder, Appl. Therm. Eng. 21(1) (2001) 53-77.
6. A.A. Al-Abidi, S. Mat, K. Sopian, M.Y. Sulaiman, A.T. Mohammad, Numerical study of PCM solidification in a triplex tube heat exchanger with internal and external fins, Appl. Therm. Eng. 61 (2013) 684-695.
7. S. Kim, L.T. Drzal, High latent heat storage and high thermal conductive phase change materials using exfoliated graphite nanoplatelets, Sol Energ Mat Sol C. 93(1) (2009) 136-142.

578447 Performance Characteristics of a Directly-Irradiated Windowless Solar Vortex Particle Receiver

We report a first-of-a-kind experimental study on the performance characteristics of a novel windowless directly-irradiated vortex-based solar particulate vessel, the Solar Expanding Vortex Receiver-Reactor (SEVR). A laboratory-scale SEVR was built and tested at a nominal capacity of 6-kWth using silicon carbide as particulate heat transfer media and a solar simulator to generate the concentrated solar radiation. A vacuum pump was used to generate a sufficiently large suction at the outlet section of the device in order to avoid particle egress through the aperture while operating the device under windowless conditions. Particle egress, flow visualisation at the aperture plane and bulk gas flow rate measurements were also carried out under iso-thermal conditions and different suction values to identify suitable conditions for carrying out hot tests. The key operating parameters, namely particle size, particle loading and gas volumetric flow rate were systematically varied to assess their influence on the thermal performance, wall cavity distribution and heat losses within the device as well as to determine the complex mechanisms controlling the behaviour of the two-phase flow in the SEVR. It was found that the improved aerodynamic of the device allows to mitigate the radiative heat losses in comparison with previous vortex-based receivers as the maximum temperature peak occurs at the opposite end to the aperture. It was also found that the the thermal performance are significantly influenced by all the parameters investigated. Particularly, the mean particle residence time, which is controlled by the particle size and gas flow rate, is a key controlling parameters influencing the maximum efficiency and temperature distribution within the device. The present investigation demonstrates for the first time that directly-irradiated vortex-based receiver technology can be efficiently operated without a window, which is the key limiting factor for scaling-up of such devices while also being a costly component subjected to damage and potential failure. The device can be employed in power generation, fuel production and high-temperature industrial processes (sensible heat, material processing) that can allow a small amount of air ingress into the receiver-reactor.

573567 Non-Thermal Equilibrium Heat Transfer Model of a Thermochemical Solar Reactor Based on Porous Media

The production of energy vectors, such as hydrogen and syngas, through thermochemical solar cycles is a promising option due to their potential high efficiencies, whose theoretical values are around 35% and 50%, respectively. To perform this process with concentrated solar energy it is necessary to develop solar reactors capable to operate with high radiative flux. An attractive reactor configuration consists on using ceramic porous media that are made or coated with the redox material. This type of reactor exhibit several advantages, which include higher volumetric yield. However, measuring temperatures inside the porous media is complicated, therefore it is necessary to develop models that allow to predict such temperatures.

In this work, a one-dimensional transient heat transfer model is proposed to evaluate the thermal behavior of a zirconium oxide foam-type porous medium exposed to concentrated solar radiation provided by the IER-UNAM Solar Furnace. The developed model considers a non-thermal equilibrium situation, in which the temperature of the porous media is different from that of the fluid. In this model an effective conductivity is used, which combines the Archie model, for the conduction contribution on the solid, and the Rosseland diffusion model, for the thermal radiation conductivity. The parameters used for the effective conductivity have been corroborated experimentally for the analyzed porous media.

The model allows calculating temperature profiles inside the porous media, which are evaluated at different values of operation parameters, such as concentrated irradiance, air flowrate, etc. The model was validated with experimental results obtained by using a 10 kW solar reactor prototype with a configuration similar to the simulated system.

Abstract 553146

ATTENUATION OF ABSORBER END LOSS IN PARABOLIC TROUGH COLLECTORS

Mohammad Abutayeh
Arkansas State University
Jonesboro, AR, USA

Kwangkook Jeong
Arkansas State University
Jonesboro, AR, USA

Anas Alazzam
Khalifa University
Abu Dhabi, UAE

Bashar El-Khasawneh
Khalifa University
Abu Dhabi, UAE

ABSTRACT

An adjustment to the conventional design of parabolic trough collectors aimed at reducing absorber end losses is suggested. The adjustment is quite simple and consists of deliberately misaligning the parabolic trough collector from its absorber in the axial direction by a predetermined amount. The structure of the parabolic trough collector remains essentially the same in the new design; however, the exposed part of the absorber will no longer be perfectly aligned with the collector due to a deliberate but calculated axial move. The cost of the new collector is expected to be equal to that of the conventional collector since the only difference between the two designs is a small misalignment. The extent of the misalignment was optimized to minimize end losses based on typical meteorological year weather data for eight locations. An Operations Research tool was employed to find the ideal level of misalignment at each location that will result in maximum incident solar radiation. A performance model of a concentrating solar power plant was then used to evaluate gains realized by the suggested misalignment at each location. The gains were modest; however, they will most likely be accomplished at almost no extra cost.

Keywords: concentrating solar power, parabolic trough collector, end loss.

INTRODUCTION

Concentrating solar power (CSP) plants are designed to focus and absorb solar heat in a solar field (SF) before forwarding it to a power block (PB) to process for electric power generation via a thermodynamic cycle. The most common method for collecting solar heat uses parabolic trough collector (PTC) assemblies to focus sunlight onto an absorber network carrying a heat transfer fluid (HTF). The HTF transfers absorbed solar heat from the SF to the PB via a heat exchanger train (HXT). A conventional PTC type CSP plant is depicted in Figure 1.

The SF of a PTC type CSP plant consists of several loops each containing a fixed number of PTC assemblies. A supply header distributes cold HTF coming from the PB among the loops and a return header collects hot HTF bound for the PB. PTC assemblies consist of numerous parabolic trough mirrors tracking the sun in an east–west rotary motion on a north–south axis as shown in Figure 2. HTF flows in an absorber pipe running in the focal line of the PTC assemblies to collect solar heat. PTC type CSP plant operation has been thoroughly described before [1]. Furthermore, proper distribution of HTF among SF loops has already been analyzed [2].

Solar heat collection in the SF is hampered by many factors like: PTC optical efficiency, mirror and absorber cleanliness, shadowing by other PTCs, loop availability, less than perfect tracking, absorber end loss, pipe heat losses, broken mirrors and glass, and other issues [3]. Improved manufacturing can raise PTC optical efficiency and tracking, while improved maintenance can increase loop availability through decreasing downtime and quickly replacing any broken mirrors or glass. Also, rapid repair of pipe and insulation ruptures will greatly minimize material and heat losses. PTC shadowing can be curtailed by increasing the row distance between SF loops; however, this will also increase the required solar collection area. The current study is focused on addressing one solvable issue: absorber end loss.

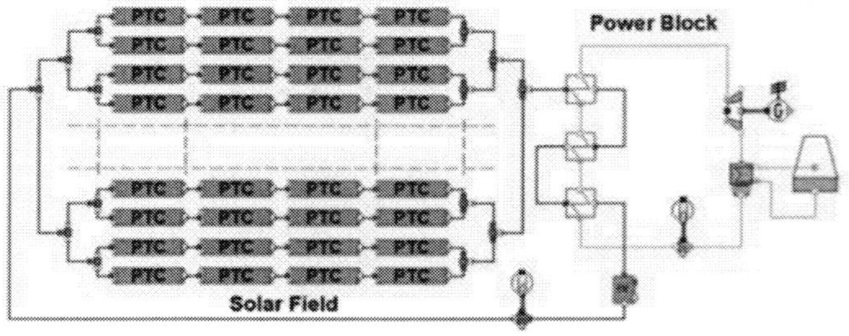

Figure 1. Conventional PTC–based CSP plant

BACKGROUND

Several researchers studied the effects of absorber end loss in PTCs on performance; however, few solutions were offered. Those few solutions involved dual axis sun tracking, which is not practical on large scale due to the linear absorber network. Even if the absorber network is somehow built to conform to the dual axis movement of the PTC, the installed cost increase would probably be intolerably significant given the large SF size. Moreover, most of those studies focused on improving the optical efficiency of PTC assemblies but overlooked the outcome on heat collection or power generation – which of course are of more significance.

A dual axis sun tracking PTC prototype was proposed [4]. The tracking is similar to conventional systems in that it tracks in an east–west rotary motion; but in addition, the entire PTC assembly was made slightly rotatable so it can align better on the north–south axis. The PTC cannot be vertically tilted, so there will always be some end loss. A small prototype was used to perform energy and exergy analysis to evaluate the advantages of this dual axis sun tracking [5]. Exergy destruction was moderately reduced and the overall exergy efficiency was increased by 3% on annual basis relative to conventional PTC assemblies.

The proposed dual axis sun tracking PTC prototype was experimentally tested with a 300 kW_{th} capacity rig [6]. The proposed PTC tracking technique reduced the solar incidence angle and end loss, consequently. Performance enhancement was felt more in the winter since the solar incidence angle is larger and therefore the end loss would have been higher in a conventional PTC assembly [7]. Also, there was little difference in performance between conventional and proposed tracking techniques in the summer due to the smaller end loss; however, this may change at different latitudes.

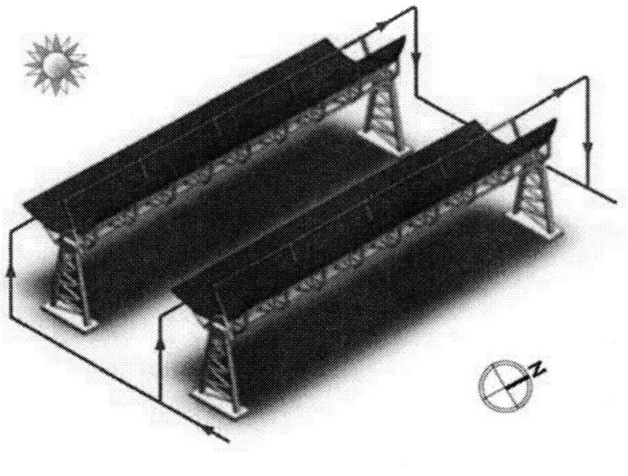

Figure 2. Conventional PTC assemblies

A similar but smaller dual axis sun tracking single PTC prototype was built and experimentally tested [8]. A computer model was also built to simulate three different tracking schemes: east–west and north–south single axis as well as dual axis. The optical efficiency and thermal performance of the PTC were notably increased when the tracking scheme was changed from single to dual axis. PTC optical efficiency was calculated based on a 24:1 concentration ratio, while PTC thermal performance was calculated based on a 200 °C HTF temperature.

A three dimensional dual axis sun tracking single PTC apparatus was built and operated using a shared sensor and software tracking mechanism [9]. Experimental data show a significant increase in heat collection compared to a fixed surface, but scaling up to industrial scale was not discussed. Another dual axis sun tracking PTC assembly was built to drive a solar thermochemical process [10]. The tracking movement was free enough to completely eliminate the end loss. A similar dual axis sun tracking PTC assembly along with its tracking control system were designed and built [11]. While PTC performance improved significantly, implementing this new PTC design in a cost effective fashion on industrial scale remains unclear.

A detailed theoretical and experimental analysis of end loss consequence on the performance of a PTC was conducted [12]. Three interesting and practically viable solutions were presented to reduce or compensate for the end loss. One of the offered solutions involve extending the exposed part of the absorber tube to capture all solar radiation reflected off the PTC. If the extension is long enough, it can totally eliminate the end loss; however, this extension will entail extra absorber cost. Another offered solution is to add a mirror at one end of the PTC structure so it can re–reflect solar radiation, that would have been otherwise lost because of the PTC end loss phenomenon, back onto the absorber. This end mirror will be attached normal to the curved mirrors of the PTC assembly which will significantly complicate mirror cleaning procedures in addition to the increased cost associated with adding a new component to the PTC structure. The end mirror will have to be located at the north end of the PTC in the northern hemisphere or at the south end of the PTC in the southern hemisphere. The last offered solution is to tilt the PTC structure to better face solar radiation. The tilt will have to be to the south in the northern hemisphere or to the north in the southern hemisphere. The ideal tilt angle that will produce the minimum average end loss is equal to the geographic latitude. The end loss will certainly be reduced by such adjustment; however, tilting is not very practical because the absorber pipe has to make turns between the PTC assemblies of the SF. This will involve extra pipe and fittings plus increased prospects for HTF leaks, which will increase installed and operating cost.

PROPOSAL

Conventional PTC assemblies track the sun by a single east–west rotary motion about a north–south axis, as shown in Figure 3. HTF flows inside an absorber pipe running in the focal line of the PTC assemblies to collect focused solar heat. This absorber pipe is exposed while running within PTC assemblies to be able to absorb reflected solar radiation, but is insulated while running between PTC assemblies to minimize heat loss from hot HTF. The exposed part of the absorber pipe is usually darkened to maximize absorptance and surrounded by a glass envelop to reduce heat loss without compromising transmittance. The exposed part of the absorber is known as a heat collection element (HCE). HCE is shown in red color while the insulated part is shown in black color in Figure 3. In conventional PTC designs, HCE is attached to PTC structures by thin supports and perfectly aligns with their mirrors. Since the sun is never fully overhead, incident and reflected sunlight will always make an acute angle with the horizontal surface of PTC mirrors known as the incident angle. Consequently, a small portion of the reflected sunlight will miss the HCE as shown in Figure 3. This sunlight reflected off the mirrors but not incident on the absorber is known as the end loss.

Altering the single axis tracking motion of the PTC to a dual axis sun tracking movement will totally eliminate the end loss; however, this is an impractical solution due to the linear absorber running in the focal line of the PTC as was mentioned before. A conceivable alternative to the dual axis tracking PTC is allow the absorber pipe to move laterally in a sliding motion in a conventional PTC design as shown in Figure 4. If the continuous lateral movement of the absorber pipe is properly controlled in such a PTC design, the end loss can be totally eliminated. This PTC design will indeed cost more than a conventional PTC design due to the nonstop sliding motion required of the absorber; therefore, it may not represent a viable option.

A practical PTC design obtained by a slight alteration of the conventional scheme is proposed. The adjustment is realized by deliberately misaligning the PTC from its absorber in the lateral direction by a fixed amount as shown in Figure 5. The size of the lateral misalignment is site specific and must be optimized for maximum incident solar radiation. The structure of the PTC and its tracking motion remain principally the same; however, the HCE will no longer perfectly align with PTC mirrors due to this lateral move. With reference to a fixed PTC, the lateral HCE shift will have to be to the north in the northern hemisphere or to the south in the southern hemisphere. The incremental cost increase associated with the new design is expected to be trivial since the only revision is a small misalignment. No extra PTC parts are required and no special tracking provisions are needed. change

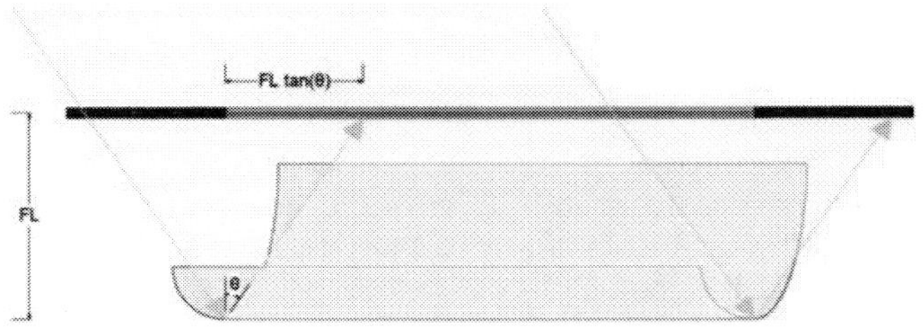

Figure 3. Conventional single axis sun tracking PTC

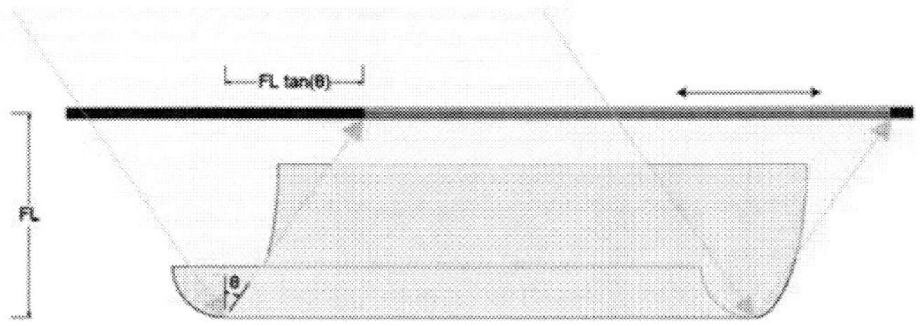

Figure 4 Optimized PTC with a sliding absorber

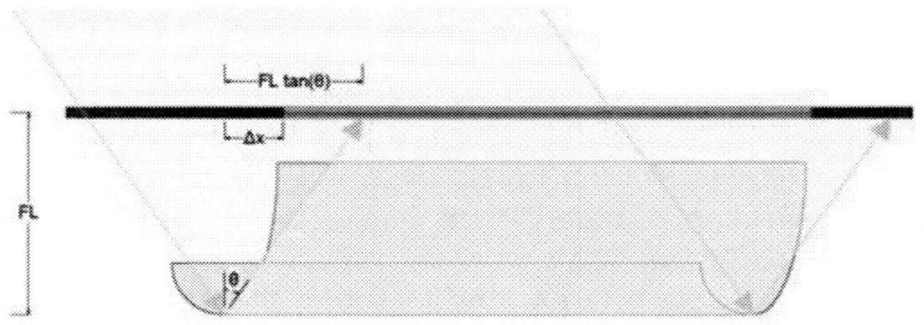

Figure 5. Proposed single axis sun tracking PTC

CALCULATION

PTC end loss calculation will be detailed next. Typical metrological year (TMY) weather data for eight different locations around the world will then be used to find the optimum PTC–absorber misalignment at each location that will minimize the impact of end loss based on the presented end loss calculation. A performance model of a 50 MW_e net rated capacity PTC type CSP plant will be run in three different PTC–absorber misalignment modes at each location: a standard mode with no misalignment as shown in Figure 3, an idealized mode with a constantly moving HCE as shown in Figure 4, and a recommended mode with the optimized PTC–absorber misalignment as shown in Figure 5. Model results for the three different PTC–absorber misalignment modes at each location will then be compared to assess the outcome of the proposed end loss attenuation scheme on solar heat collection, electric power generation, and CSP plant revenue.

The ensuing equations were disclosed in earlier work concerning CSP plant simulation [1] and are taken from research findings done at Sandia National Laboratories [13] [14]. They are used to compute solar time to then compute solar angles. Next, attenuation of solar radiation is then estimated to quantify captured solar heat.

TC is a time correction term needed to adjust regular time to solar time. TC is a function of time zone, longitude angle, and day of year and is estimated by

$$TC = 9.87 \cdot sin(2x) - 7.53 \cdot cos(x) - 1.5 \cdot sin(x) - 4 \cdot (15\,TZ - l) \\ x = 2\pi \cdot (Day - 81)/365 \tag{1}$$

SD is the solar day which is also needed to adjust regular time to solar time and is given by

$$SD = \begin{cases} Day, & 0 \leq Hour + \frac{TC}{60} \leq 24 \\ Day - 1, & Hour + \frac{TC}{60} < 0\ AND\ Day > 1 \\ Day + 364, & Hour + \frac{TC}{60} < 0\ AND\ Day \leq 1 \\ Day + 1, & Hour + \frac{TC}{60} > 24\ AND\ Day < 365 \\ 1, & Hour + \frac{TC}{60} > 24\ AND\ Day \geq 365 \end{cases} \tag{2}$$

SH is the solar hour used later to calculate several solar angles and is given by

$$SH = \begin{cases} Hour + \frac{TC}{60}, & 0 \leq Hour + \frac{TC}{60} \leq 24 \\ 24 + Hour + \frac{TC}{60}, & Hour + \frac{TC}{60} < 0 \\ Hour + \frac{TC}{60} - 24, & Hour + \frac{TC}{60} > 24 \end{cases} \tag{3}$$

Next, declination, hour, altitude, azimuth, and altitude transverse solar angles are given by

$$\delta = \sin^{-1}\left[0.39795 \cdot \cos\left(0.98563 \cdot \frac{\pi}{180} \cdot [SD - 173]\right)\right] \tag{4}$$

$$HA = \frac{\pi}{180} \cdot 15 \cdot (SH - 12) \tag{5}$$

$$\alpha = \sin^{-1}[\sin(\delta) \cdot \sin(\lambda)] + \cos(\delta) \cdot \cos(HA) \cdot \cos(\lambda) \tag{6}$$

$$a = \frac{\sin(\alpha) \cdot \sin(\lambda) - \sin(\delta)}{\cos(\alpha) \cdot \cos(\lambda)} \tag{7}$$

$$AT = \begin{cases} \alpha, & |a - \pi| < 1 \\ \tan^{-1}\left(\frac{\tan(\alpha)}{\left|\cos\left(\frac{\pi}{2} + a\right)\right|}\right), & |a - \pi| \geq 1 \end{cases} \tag{8}$$

SA is the shadow argument needed to evaluate shadow effects on incident solar radiation. SA is a function of PTC row distance and aperture width plus the altitude transverse solar angle

$$SA = \frac{RD}{AW} \cdot \cos\left(\frac{\pi}{2} - AT\right) \tag{9}$$

Shadow efficiency is a multiplier used in adjusting direct incident solar radiation to account for PTC shadow eclipsing the SF around sunrise and sunset

$$\eta_{Shadow} = \begin{cases} 1, & SA \geq 1 \\ 0, & SA < 0 \\ SA, & 0 \leq SA < 1 \end{cases} \tag{10}$$

IA is the incident angle defined as the angle between solar beams and the line normal to the PTC aperture. It is constantly changing and can be calculated by

$$IA = \cos^{-1}\left\{\sqrt{1 - [\cos(\alpha) \cdot \cos(a)]^2}\right\} \tag{11}$$

The above IA formula is appropriate at 0° PTC tilt angle (horizontally levelled PTC) and 0° orientation angle (north–south positioned PTC axis). Otherwise, tilt and orientation angles should be considered in the above IA calculation as was described in previous work [1].

IAM is the incident angle modifier used to adjust incident solar radiation to account for direct and indirect losses due to incident angle. IAM is a multiplier that can be estimated by the following correlation

$$IAM = \cos(IA) - 0.03008 \cdot IA - 0.09389 \cdot IA^2 \tag{12}$$

End loss efficiency is a multiplier used for adjusting incident solar radiation to account for end losses, solar radiation incident on PTC mirrors but not landing on PTC absorber or HCE. End loss efficiency is given by

$$\eta_{End\,Loss} = 1 - \frac{|FL \cdot \tan(IA) - \Delta x|}{L_{PTC}} \tag{13}$$

The term *I FL· tan (IA) – Δx I* represents the extent of HCE missed by the concentrated and reflected incident solar radiation as illustrated by Figure 3, Figure 4, and Figure 5. The quantity *Δx* represents the extent of the PTC–absorber lateral misalignment; thus, there are three possibilities for its calculation: *Δx = 0* in a standard PTC design as shown in Figure 3 signifying no misalignment, *Δx = FL· tan (IA)* in an ideal PTC design with a sliding absorber shown in Figure 4 signifying no end loss due to continual tracking, or *Δx = B* in the proposed PTC design shown in Figure 5 signifying a fixed misalignment.

Accordingly, the quantity Δx can be expressed by

$$\Delta x = \begin{cases} 0, & PTC\ Misalignment = Normal \\ B, & PTC\ Misalignment = Fixed \\ FL\tan(IA), & PTC\ Misalignment = Tracking \end{cases} \quad (14)$$

The constant B in the above equation is the proposed PTC–absorber misalignment. Its value will be different at different locations as adjusted by an operations research tool to maximize incident solar radiation.

Absolute efficiency is an overall multiplier used to compute the incident part of available solar radiation. In the context of a PTC based CSP plant, it combines the effects of optical efficiency, mirror cleanliness, shadowing, incident angle, percent of PTC assemblies in service, and end loss [15]. Absolute efficiency, $\eta_{Absolute}$, is given by

$$\eta_{Absolute} = \eta_{Optical} \cdot \eta_{Shadow} \cdot \eta_{End\ Loss} \cdot C_{Mirror} \cdot IAM \quad (15)$$

Direct normal irradiance (DNI) is beam solar radiation available for exploitation, while direct incident irradiance (DII) is the slice of DNI that is incident on PTC absorbers. DII is given by

$$DII = \eta_{Absolute} \cdot DNI \quad (16)$$

Solar times and angles in addition to DNI attenuation efficiencies were calculated using the above equations at eight different locations spatially spread over the globe. The calculations were performed in one hour increments for a whole year to match TMY data format. TMY weather data at those eight different locations were obtained from the National Renewable Energy Laboratory (NREL) web site. The same standard *EuroTrough* design was used for PTC assemblies at all locations; furthermore, the same generic values of optical efficiency, mirror cleanliness, and PTC availability were used at all locations.

The above equations were coded in *Microsoft Excel*. *Solver* is a *Microsoft Excel* add–in that can be used to perform operations research type analysis on large data sets. *Solver* was used to find the optimum PTC–absorber misalignment extent, B, at each site that will maximize the sum of incident solar radiation over the TMY period of each location: $\Sigma\ (DII \cdot \Delta t)_{Maximum} = f\ (B_{Optimum})$.

A performance model of a 50 MW_e net rated capacity PTC type CSP plant with a properly sized thermal energy storage (TES) system will be run in the aforementioned PTC–absorber misalignment modes. Model results of captured solar heat and power generation for the three different PTC–absorber misalignment modes at each location will then be compared to assess the outcome of the proposed end loss attenuation scheme on solar heat.

RESULTS

End loss is constantly changing depending on solar times and angles. Figure 6 illustrates four daily variations of end loss at an undisclosed location in the northern hemisphere, hence the hidden y–axis values. Since solar radiation is experienced between sunrise and sunset, end loss and other solar calculations before sunrise or after sunset are irrelevant. End losses are least in the summer and most in the winter during the daytime operation.

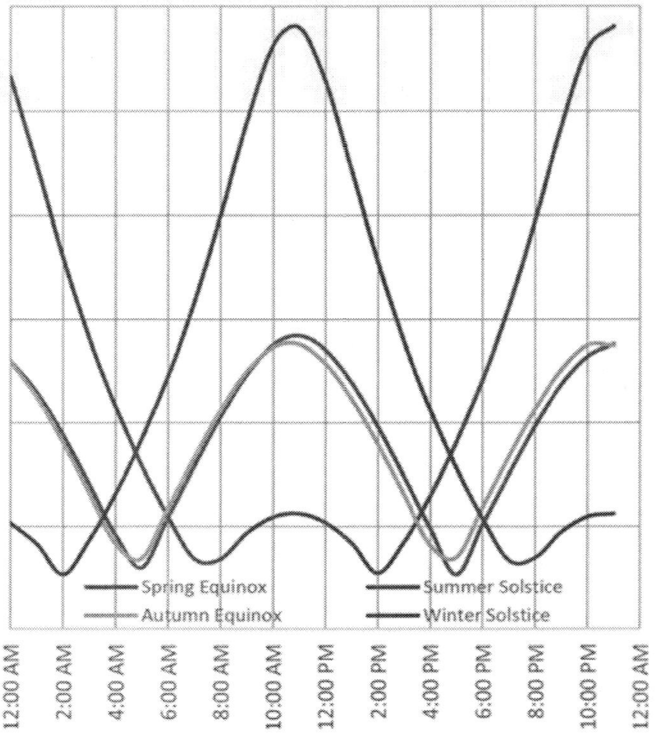

Figure 6. Daily variation of end loss at a specific location

The end loss is constantly changing depending on solar times and angles as mentioned above; however, the extent of that end loss depends on location, namely the latitude angle. The following polynomial curve fit was obtained by regressing end loss data at different latitudes

$$L^{Maximum}_{End\ Loss} = a + b\,\lambda + c\,\lambda^2 + d\,\lambda^3 + e\,\lambda^4 + f\,\lambda^5 + g\,\lambda^6 \quad (17)$$

Where a = 7.87873E-01, b = -9.18497E-05, c = 2.75105E-03, d = 1.70851E-07, e = -1.53553E-06, f = -6.15529E-11, g = 4.91693E-10

The polynomial expresses the maximum possible end loss as a function of the latitude angle. Figure 7 plots the previous polynomial and shows that end losses increase significantly with latitude in both hemispheres.

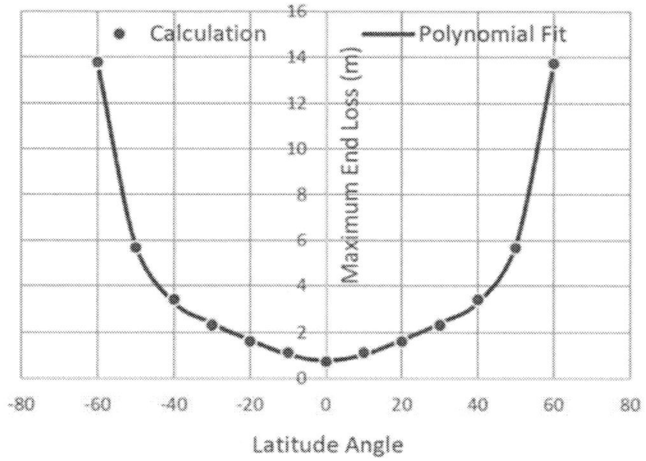

Figure 7. Global maximum end loss versus latitude

In addition, Figure 8 demonstrates the daily variation of end loss on an arbitrary day at different latitudes. The arbitrary day taken to show the daily variation was spring equinox selected for its average outcomes. The absolute value of the latitude angle was used to generalize results for both hemispheres. It is evident from Figure 8 that end loss increases with latitude as mentioned before.

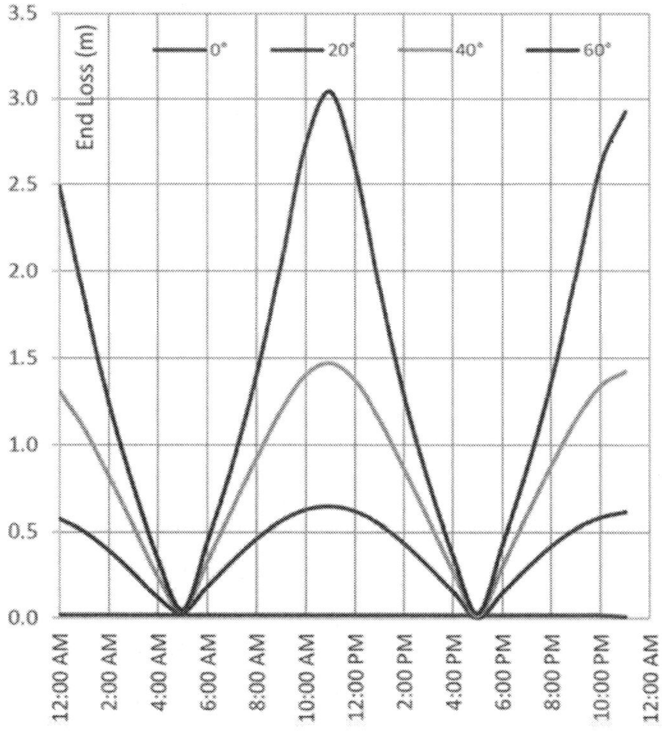

Figure 8. Spring equinox end loss trends at different $|\lambda|$

A performance model of a 50 MW_e net rated capacity PTC type CSP plant fitted with a TES system will be used to evaluate gains realized by implementing the new PTC design. CSP performance modeling has been described in detail before [1] so has TES operation and control [16]. Transient operation is the norm in CSP plants; therefore, previously described model modifications were adapted in this performance model to properly account for transient start up and shut down operations as well as cloud cover events and HTF residence time in the SF [17]. The model was run to simulate an annual operation at eight different locations around the world shown in Table 1. The model was not altered to optimize the size of the SF with respect to available DNI at each location. Thus, the eight different simulations differed in only four items: the three constants shown in Table 1 and DNI data.

Table 1: Selected CSP plant locations

Location	*λ (°)*	*l (°)*	*TZ (h)*
Adelaide, Australia	-34.93	138.52	9.50
Antofagasta, Chile	-23.43	-70.43	-4.00
Delingha, China	37.37	97.37	8.00
Aswan, Egypt	23.97	32.78	2.00
Bikaner, India	28.05	73.35	5.50
Casablanca, Morocco	33.37	-7.58	0.00
Sevilla, Spain	37.42	-5.90	1.00
Palm Springs, USA	33.83	-116.50	-8.00

The above equations were coded in *Microsoft Excel*, where the *GRG Nonlinear* engine of its *Solver* add–in was used to find the optimum PTC–absorber misalignment, *B*, that will maximize total annual DII at each location based on its TMY records: $\Sigma\ (DII \cdot \Delta t)_{Maximum} = f\ (B_{Optimum})$. Table 2 lists total DNI taken from the TMY weather files of NREL as well as the maximum reached end loss and optimum PTC–absorber misalignment found at each location.

Table 2: Optimum PTC–absorber misalignment found

Location	*Σ DNI (kWh/m²)*	$L^{Maximum}_{End\ Loss}$ *(m)*	*B (m)*
Adelaide, Australia	1,691	2.76	0.48
Antofagasta, Chile	2,044	1.82	0.37
Delingha, China	1,544	3.00	0.79
Aswan, Egypt	2,434	1.86	0.40
Bikaner, India	1,739	2.14	0.58
Casablanca, Morocco	1,583	2.58	0.48
Sevilla, Spain	1,773	3.03	0.57
Palm Springs, USA	2,534	2.65	0.50

DNI normally decline with latitude as can be inferred from TMY files or other empirical solar radiation models; nonetheless, clouds, humidity, dust, aerosols, and other uncontrolled natural phenomena can result in deviation from this general rule. Totalized DNI records taken from TMY files shown in Table 2 validate this principle. It is also evident that the maximum reached end loss and optimum PTC–absorber misalignment increase with latitude as can be seen in Table 2. It is imperative to restate that optimum PTC–absorber misalignment results listed in Table 2 were obtained based on TMY data which should be sufficient to show the benefits of the proposed misalignment; though, actual DNI and other weather conditions vary significantly from TMY data on everyday operation. Therefore, it may be a better practice to optimize the size of the proposed PTC–absorber misalignment based on DNI records from an empirical cloudless model rather than TMY data.

Next, the aforementioned performance model of a 50 MW_e net rated capacity PTC type CSP plant fitted with an adequately sized TES system will be run in three different PTC–absorber misalignment modes to assess the effect of the proposed new PTC design on solar heat collection, electric power generation, and plant revenues. The three PTC–absorber misalignment modes are: standard mode with no misalignment shown in Figure 3, idealized mode with a constantly moving HCE shown in Figure 4, and the recommended mode with the optimized PTC–absorber misalignment shown in Figure 5. Then, annual simulation results for the three different PTC–absorber misalignment modes at each location will be compared to assess the outcome of the proposed end loss attenuation scheme. At each location, results are presented in three categories: *Normal* refers to the standard mode with no misalignment producing the worst results, *Fixed* refers to the proposed PTC–absorber misalignment mode, and *Tracking* refers to the idealized mode with a constantly moving HCE in the lateral direction yielding the best possible results.

Table 3 shows the annual model output of solar heat forwarded to the PB for direct electric power generation. The amount of heat transfer slightly increased when the proposed PTC–absorber misalignment was implemented by 0.1 – 0.4%. The amount of heat transfer can increase by 0.2 – 1.0% if a continually tracking HCE is used. This slight increase is due to having an oversized SF designed to simultaneously meet the heat demand of the PB and thermally charge the TES system. Table 4 shows the annual model output of excess solar heat sent to the TES system for later reclamation. The amount of heat transfer modestly increased when the proposed PTC–absorber misalignment was used by 0.3 – 1.1%. The amount of heat transfer can increase by 0.9 – 2.4% if a continually tracking HCE is used. The above results will be different if different PTC dimensions were used.

Table 3: Heat transfer from SF to PB, GWh_{th}

Location	*Normal*	*Fixed*	*Tracking*
Adelaide, Australia	296.14	296.46	298.13
Antofagasta, Chile	391.16	391.37	391.92
Delingha, China	262.41	263.53	264.98
Aswan, Egypt	465.50	465.87	466.38
Bikaner, India	399.47	399.72	400.75
Casablanca, Morocco	294.48	294.69	295.48
Sevilla, Spain	331.07	331.38	332.91
Palm Springs, USA	458.44	459.15	459.69

Table 4: Heat transfer from SF to TES, GWh_{th}

Location	*Normal*	*Fixed*	*Tracking*
Adelaide, Australia	116.14	116.57	117.13
Antofagasta, Chile	178.00	178.55	179.64
Delingha, China	50.42	50.97	51.64
Aswan, Egypt	233.49	234.13	235.62
Bikaner, India	65.55	66.08	66.75
Casablanca, Morocco	97.02	97.56	98.15
Sevilla, Spain	106.36	106.95	107.66
Palm Springs, USA	194.55	195.79	196.70

If the results of Table 3 and Table 4 are combined, the amount of solar heat collected in the SF can increase by 0.1 – 0.5% if the proposed PTC–absorber misalignment is used and by 0.4 – 1.2% if a continually tracking HCE is used. With reference to location, least improvement was realized at Antofagasta, Chile while most improvement was realized at Delingha, China.

Table 5 presents the annual model output of electric power generation. The amount of electricity generated modestly increased when the proposed PTC–absorber misalignment was applied by 0.1 – 0.5%. The amount of electric power generation can increase by 0.4 – 1.2% if a continually tracking HCE is used. With regard to location, least growth was realized at Antofagasta, Chile while most growth was realized at Delingha, China.

Table 6 shows the annual model output of CSP plant revenues based on a 50 $/$MWh_e$ net renewable electricity value. Revenues modestly increased when the proposed PTC–absorber misalignment was applied by 0.1 – 0.5% matching the above generation results. Revenues can increase by 0.4 – 1.2% if a continually tracking HCE is used matching the above generation results. With regard to location, least growth was realized at Antofagasta, Chile while most growth was realized at Delingha, China.

Table 5: Gross power generation, GWh_e

Location	*Normal*	*Fixed*	*Tracking*
Adelaide, Australia	153.18	153.45	154.28
Antofagasta, Chile	212.58	212.86	213.51
Delingha, China	114.07	114.66	115.39
Aswan, Egypt	262.28	262.64	263.41
Bikaner, India	171.04	171.36	172.00
Casablanca, Morocco	145.10	145.39	145.98
Sevilla, Spain	162.74	163.13	163.95
Palm Springs, USA	244.30	244.81	245.68

Table 6: CSP plant revenues, $

Location	*Normal*	*Fixed*	*Tracking*
Adelaide, Australia	$6.89 M	$6.91 M	$6.94 M
Antofagasta, Chile	$9.57 M	$9.58 M	$9.61 M
Delingha, China	$5.13 M	$5.16 M	$5.19 M
Aswan, Egypt	$11.8 M	$11.82 M	$11.85 M
Bikaner, India	$7.7 M	$7.71 M	$7.74 M
Casablanca, Morocco	$6.53 M	$6.54 M	$6.57 M
Sevilla, Spain	$7.32 M	$7.34 M	$7.38 M
Palm Springs, USA	$10.99 M	$11.02 M	$11.06 M

CONCLUSION

A modification of the standard PTC design intended to reduce HCE end losses is suggested. The modification consists of purposefully misaligning PTC from its HCE absorber in the axial direction by a preset amount. The structure of the PTC remains essentially the same in the newly proposed design; however, the HCE will no longer precisely align with the collector due to the deliberate but calculated axial misalignment. The cost of the new PTC is expected to be equal to that of a typical PTC since the only difference between the two PTC designs is a small axial PTC–absorber misalignment.

The size of the proposed misalignment was optimized to minimize HCE end losses based TMY weather data at eight international sites. *Solver*, an optimization engine available as a *Microsoft Excel* add–in, was employed to find the ideal misalignment magnitude at each location that will cause maximum incident solar radiation. A performance model of a PTC type CSP plant was then utilized to evaluate the outcome of the proposed axial PTC–absorber misalignment on solar heat collection, electric power generation, and plant revenues. The gains varied with location but were generally modest; even so, they will be likely attained at no extra cost.

The US Department of Energy launched the *SunShot* Initiative in early 2011 with the aim of reducing the cost of employing solar energy to levels equivalent to the cost of employing conventional energy by the end of the decade. The current exploration is another attempt in a series of attempts intended for increasing the viability of exploiting solar energy. More enhanced CSP modeling techniques were disclosed before [1] [17]; moreover, CSP operating procedures to manage HTF flow in SF [2] and control the charging and discharging processes of TES systems [16] were suggested to improve operation and upturn profits.

Also, a scheme to streamline the usually oscillating power output of CSP plants under transient conditions was proposed [18] and currently pending patent approval [19]. The proposed scheme slightly decouples solar heat supply from sink allowing for more control of heat transfer rates thus smoothing out electric power generation. The proposed scheme aims to solve the problem of electricity oscillations usually associated with the intermittent solar energy thus tolerating more contribution from CSP plants to the electric grid. Moreover, electric power generation slightly increased under the proposed scheme because the PB ran at a slightly higher thermal efficiency due to better dispersion of solar heat to the PB.

Doping SF HTF with nanoparticles was also explored [20] but was found to be practically useless since it had almost no effect on power generation not to mention the extra cost of nanoparticles and arising new operational challenges like sedimentation and abrasion. The current study falls within the same theme of reducing the cost of CSP deployment to make it more feasible as it aims to attenuate the negative impact of the inevitable absorber end loss. Meeting the target of the *SunShot* Initiative will most likely transpire by a number of minor incremental improvements rather than a single silver bullet.

Other potential prospects to optimize CSP operation to meet the target of the *SunShot* Initiative may involve altering the tilt and orientation angles of PTC assemblies. adjusting the tilt angle is very challenging as it will involve sloping the entire SF, but adjusting the orientation angle, polar rather than north–south positioned PTC axis, seems plausible. Pressurized HTFs currently used in PTC type CSP plants boil around 400 °C. If higher boiling HTF were to be developed, the PB working fluid can reach higher temperatures and subsequently raising thermal efficiency. Incorporating a TES system in CSP operation decouples heat supply from sink a bit thus raising its dispatchability. Operation Research tools can then be used to optimize CSP dispatch to maximize revenues based on a validated performance and financial plant model. Lastly, integrating other operations with CSP, like desalination and chemical processing, can synergistically increase feasibility.

NOMENCLATURE

a	azimuth angle, radians
AT	altitude transverse, radians
AW	aperture width, m
B	optimized misalignment
C	mirror cleanliness, percent
CSP	concentrating solar power
Day	day of year, day
DII	direct incident insolation, W/m²
DNI	direct normal insolation, W/m²
FL	focal length, m
HA	hour angle, radians
HCE	heat collection element
Hour	hour of day, hour
HTF	heat transfer fluid
HXT	heat exchanger train
IA	incident angle, radians
IAM	incident angle modifier, percent
L	length, m
l	longitude angle, degrees
NREL	National Renewable Energy Laboratory
PB	power block
PTC	parabolic trough collector
RD	row distance, m
SA	shadow argument
SD	solar day, day
SF	solar field
SH	solar hour, hour
TC	time correction term, minute
TES	thermal energy storage
TMY	typical metrological year
TZ	time zone, hour
α	altitude angle, radians
δ	declination angle, radians
Δx	PTC–absorber lateral misalignment
η	efficiency, percent
λ	latitude angle, degrees

REFERENCES

[1] M Abutayeh, DY Goswami, EK Stefanakos. Solar Thermal Power Plant Simulation. Environmental Progress & Sustainable Energy Journal. July 2013.

[2] M Abutayeh, A Alazzam, B Khasawneh. Balancing Heat Transfer Fluid Flow in Solar Fields. Solar Energy Journal. July 2014.

[3] S Al-Hanaei, S Al-Shomali, M Abutayeh. Performance Model of Shams I Solar Power Plant. Heat Transfer Engineering Journal. January 2016.

[4] Jie Sun, Ruilin Wang, Hui Hong, Qibin Liu, An optimized tracking strategy for small-scale double-axis parabolic trough collector, Applied Thermal Engineering, Volume 112, 2017, Pages 1408-1420, ISSN 1359-4311.

[5] Ruilin Wang, Wanjun Qu, Hui Hong, Jie Sun, Hongguang Jin, Experimental performance of 300 kWth prototype of parabolic trough collector with rotatable axis and irreversibility analysis, Energy, Volume 161, 2018, Pages 595-609, ISSN 0360-5442.

[6] Wanjun Qu, Ruilin Wang, Hui Hong, Jie Sun, Hongguang Jin, Test of a solar parabolic trough collector with rotatable axis tracking, Applied Energy, Volume 207, 2017, Pages 7-17, ISSN 0306-2619.

[7] Wanjun Qu, Ruilin Wang, Hui Hong, Jie Sun, Hongguang Jin, Prototype Testing of a 300kWth Solar Parabolic-trough Collector Using Rotatable Axis Tracking, Energy Procedia, Volume 105, 2017, Pages 780-786, ISSN 1876-6102.

[8] Fahim, U. and M. Kang "Performance Evaluation of Dual-axis Tracking System of Parabolic Trough Solar Collector." IOP Conference Series: Materials Science and Engineering 301(1): 012166.

[9] George C. Bakos, Design and construction of a two-axis Sun tracking system for parabolic trough collector (PTC) efficiency improvement, Renewable Energy, Volume 31, Issue 15, 2006, Pages 2411-2421, ISSN 0960-1481.

[10] Juan Fang, Qibin Liu, Taixiu Liu, Jing Lei, Hongguang Jin, Thermodynamic evaluation of a distributed energy system integrating a solar thermochemical process with a double-axis tracking parabolic trough collector, Applied Thermal Engineering, 2018, ISSN 1359-4311.

[11] A. E. Gaafar and A. F. Zobaa, "Economical design of a two-axis tracking system for solar collectors," 5th IET International Conference on Renewable Power Generation (RPG) 2016, London, 2016, pp. 1-6.

[12] Ming Li, Chengmu Xu, Xu Ji, Peng Zhang, Qiongfen Yu, A new study on the end loss effect for parabolic trough solar collectors, Energy, Volume 82, 2015, Pages 382-394, ISSN 0360-5442.

[13] Lippke F. Simulation of the Part Load Behavior of a 30MWe SEGS Plant. SAND95-1293. Sandia National Laboratories; Albuquerque, NM: 1995

[14] Cohen GE, Kearney DW, Kolb GJ. Final Report on the Operation and Maintenance Improvement Program for Concentrating Solar Power Plants.

SAND99-1290. Sandia National Laboratories; Livermore, CA: 1999.

[15] Rheinländer J, Bergmann S, Erbes MR. Technical and Economic Performance of Parabolic Trough Solar Power Plants – A Computational Tool for Plant Feasibility Studies. 14th SolarPACES International Symposium on Concentrated Solar Power and Chemical Energy Technologies; Las Vegas, NV: 2008.

[16] M Abutayeh, A Alazzam, B Khasawneh, Optimizing Thermal Energy Storage Operation, Solar Energy Journal, October 2015.

[17] M Abutayeh, A. Alazzam, Adapting Steady-State Solar Power Models to Include Transients. Solar Energy Engineering Journal, April 2017.

[18] M Abutayeh, A Alazzam, B Khasawneh, Streamlining the Power Generation Profile of Concentrating Solar Power Plants, ASME 12th International Conference on Energy Sustainability, Lake Buena Vista, FL, June 2018

[19] US Patent Pending: System And Method To Stabilize Power Generation Of Concentrated Solar Power Plants, Filed in July 2017.

[20] M Abutayeh, Y Addad, E Abu–Nada, A Alazzam, Doping Solar Field Heat Transfer Fluid with NanoParticles, Solar Energy Engineering Journal, February 2019.

CITATION

M Abutayeh, K Jeong A Alazzam, B El-Khasawneh, *Attenuation of Absorber End Loss in Parabolic Trough Collectors*, ASME 2019 13th International Conference on Energy Sustainability, Bellevue, WA, July 2019

574593 Modal Switching for Continuous Solar Reforming of Methane

The carbon footprint of methane reforming processes can be reduced by using solar thermal energy in place of fossil energy for process heating. In this scheme, concentrated solar power provides heat that drives endothermic reformation of methane, storing renewable energy in the chemical bonds of the resultant synthesis gas. CO_2 can be directly valorized in this process to further offset end-use carbon emissions. However, solar transience hinders the operational window and overall efficiency of these systems, limiting their practical feasibility. Here predictive switching between bi-reforming and tri-reforming of methane modes over a single nickel-based catalyst is applied in order to operate solar methane reforming continuously. This is performed using Oregon State University's solar simulator and directly-irradiated volumetric reactor, which is designed to promote fast transient behavior of solar-thermochemical reactions. Solar-to-chemical efficiency is calculated over periods of transient modal switching, where solar power is reduced incrementally and oxygen with additional methane are throttled into the feed to compensate for this decrease in external heating. Catalyst stability and coke formation are examined after modal switching.

567405 Sustainable Steam on Demand for the Bayer Process

Sustainable Steam on Demand for the Bayer Process

Alicia Bayon and Andrew Beath

CSIRO Energy, PO Box P.O. Box 330, Newcastle, NSW 2300, Australia

In this work, we explore the coupling of Concentrated Solar Thermal (CST) to produce the steam required for the Bayer Process used for the conversion of bauxite to aluminium hydroxide. Specifically, we present the use parabolic trough and central receiver tower technologies to produce the required steam for the digestion and electricity demand of a commercial-scale Bayer Process plant (392 MW_{th}). This work explores the integration of different configurations for the production of steam using annual simulations with real direct normal irradiation data. The effect of the solar field size as a critical parameter towards the economic optimisation of the plant is investigated. In addition, we analyzed plausible sites based on physical and operational constraints. We fixed two targets, a minimum CST capacity factor of 29% and a minimum NPV of $20 million. The internal rate of return (IRR) is also used to find the optimum design conditions at a specific location.

Results for trough systems are shown as an example at Learmonth. Other sites in Australia like Gladstone, Pinjarra and Darwin (Australia) are also explored. In this abstract, Learmonth is shown as an example of achieving the target CST input of minimum 29% at relatively small field and small storage sizes (see white line in Figure 1.a), and achieving positive net present value (NPV) results that increase as the system size increases (see background in Figure 1.a). Other sites like Gladstone also achieves good results, with a local NPV maximum at solar multiple 2 with 2 hours of storage providing a CST input of approximately 34%. Darwin, does not appear to be as promising for the technology and negative NPVs result from all combinations meeting the target CST input. Selection of the optimum at Learmonth is limited by the maximum IRR obtained which occurs at solar multiple 2 with 6 hours of storage. The IRR values are over 12% in most of the design conditions for Learmonth and Gladstone. The maximum is localized in solar multiple 2 and 6 hours storage for Learmonth where the IRR is 17.0% (Figure 1.b).

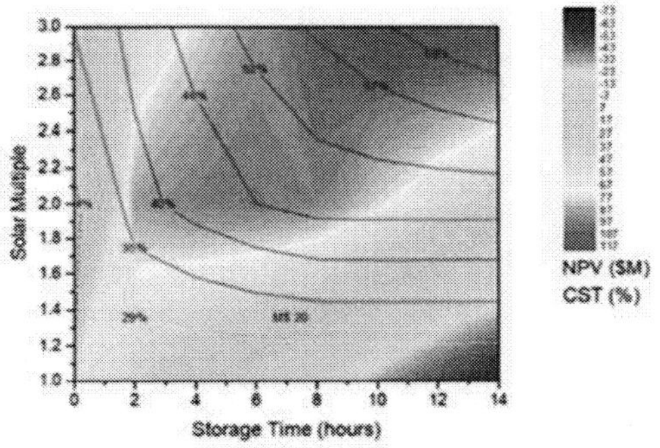

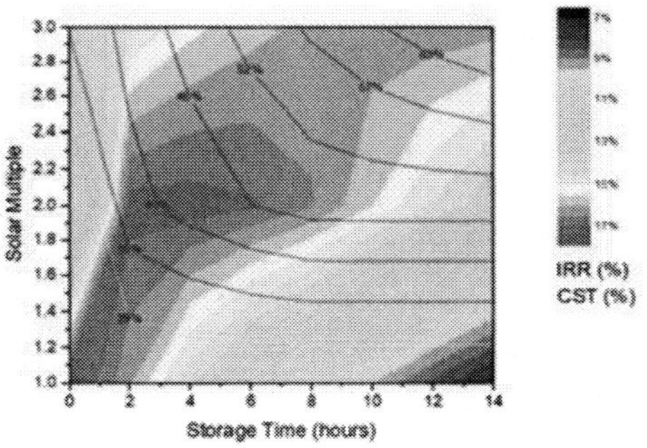

Figure 1: Net present value (NPV), internal rate of return (IRR) and capacity factor (CST) of parabolic trough plants (392 MW_{th}) using the cost model at Learmonth.

560193 In-Situ Electrochemical Techniques to Determine Ionomer Coverage in PEFC Electrodes

Many electrochemical devices such as low temperature fuels utilize ionomers to improve efficiency and overall performance. In polymer electrolyte fuel cells (PEFCs), perflurosulfonic acid ionomers improve ionic (protonic) conductivity and Pt utilization within the catalyst layer. However, the strong interaction between sulfonate ($-SO_3^-$) functional groups and Pt nanoparticles have been linked lower catalytic rates and higher O_2 transport resistances.[1,2] There have been numerous strategies implemented to control the local distribution of ionomer and Pt/C catalyst particles, but limitations in the resolution of state-of-the-art characterization techniques make it difficult to assess how ink formulations and/or electrode fabrication techniques control ionomer-catalyst interactions and its impact device performance.[3–5]

Herein, the development and utilization of *in situ* electrochemical techniques designed to probe the local ionomer interactions will be utilized in conjunction with established testing protocols to relate ionomer coverage (on Pt and/or C support) with kinetic performance and O_2 transport resistance of fully-conditioned PEFC electrodes. Specifically, CO displacement chronoamperometry and electrochemical impedance spectroscopy techniques will be described to determine local ionomer coverages across a series of Pt/C MEAs. These results will promote a deeper understanding of ionomer-catalyst interactions which in turn will inform future materials development and integration necessary to achieve high performance PGM (e.g. Pt/C and Pt alloy/C) electrocatalyst-based electrodes.

References

1. Schuler, T. *et al.* Fuel-Cell Catalyst-Layer Resistance via Hydrogen Limiting-Current Measurements. *J. Electrochem. Soc.* **166**, F3020–F3031 (2019).
2. Subbaraman, R., Strmcnik, D., Paulikas, A. P., Stamenkovic, V. R. & Markovic, N. M. Oxygen Reduction Reaction at Three-Phase Interfaces. *ChemPhysChem* **11**, 2825–2833 (2010).
3. Berlinger, S. A., McCloskey, B. D. & Weber, A. Z. Inherent Acidity of Perfluorosulfonic Acid Ionomer Dispersions and Implications for Ink Aggregation. *J. Phys. Chem. B* **122**, 7790–7796 (2018).
4. Orfanidi, A., Rheinländer, P. J., Schulte, N. & Gasteiger, H. A. Ink Solvent Dependence of the Ionomer Distribution in the Catalyst Layer of a PEMFC. *J. Electrochem. Soc.* (2018). doi:10.1149/2.1251814jes
5. Khandavalli, S. *et al.* Rheological Investigation on the Microstructure of Fuel Cell Catalyst Inks. *ACS Appl. Mater. Interfaces* (2018). doi:10.1021/acsami.8b15039

577465 Overpotential Analysis of Low Loading Pt/C Electrodes Developed By Reactive Spray Deposition Technology

According to the United States Environmental Protection Agency, 28% of greenhouse gas emitted in the United States is the result of transportation [1]. Lately, there has been a significant focus from industry to develop green technology for transportation. In particular, hydrogen fuel cell vehicles have become a major focus. As fuel cell and electrolyzer technologies become more mature and more hydrogen fueling stations are constructed nationwide, fuel cell vehicles can become a more viable green transportation method.

However, for fuel cell vehicles to become a viable transportation option in the future, additional research is required. Fuel cells will need to be optimized for the low humidity operating conditions required for automotive applications. One method to achieve this is to adjust the ionomer used in electrode fabrication. With these parameters that need to be improved, additional research should be performed to further decrease the costs associated with fuel cell manufacturing. One such fuel cell fabrication method is Reactive Spray Deposition Technology (RSDT). RSDT is a one-step flame-based process which produces platinum from the primary flame from a precursor solution while simultaneously incorporating a carbon/ionomer slurry from secondary spray nozzles. The resulting Pt/C mix is directly deposited onto proton exchange membranes to create catalyst-coated membranes (CCMs) without an additional drying step. Previous work by Yu et al examined the use of RSDT to create low Pt loading electrodes for fuel cell applications [2].

In order to improve the fuel cell performance at low humidity conditions, it is important to examine a change in ionomer. In traditional proton exchange membrane (PEM) fuel cells, it is common to use a Nafion membrane and Nafion ionomer. However, shortening the side chain length from standard Nafion 1100EW ionomers has shown to have the potential to improve the fuel cell performance at low relative humidity. This work will examine various equivalent weight Nafion and Solvay ionomers and their effect on fuel cell performance at low humidity.

The work presented will use the RSDT process to fabricate Pt/C electrodes with baseline materials as well as varying ionomer intended to improve the performance of the fuel cell at low humidity conditions. In addition, this work will look to analyze the electrochemical data by utilizing the six-step overpotential analysis technique used by Yu et al to examine how these changes affect the contributions from the different polarization sources [2, 3]. We will also look to compare the results of this overpotential analysis with the distribution of relaxation times information determined from electrochemical impedance spectroscopy in order to get a more complete understanding of how the changes in ionomer will impact the fuel cell performance, where the voltage loss occurs and how to further optimize the fuel cells for automotive applications.

References

[1] U.S. EPA. "Sources of Greenhouse Gas Emissions" www.epa.gov/ghgemissions/sources-greenhouse-gas-emissions

[2] H. Yu, L. Bonville, R. Maric. *Journal of the Electrochemical Society*, **165**, 5 (2018).

[3] H. Yu, L. Bonville, R. Maric. *Journal of the Electrochemical Society*, **165**, 15 (2018).

564743 Rechargeable Aluminum-Graphite Batteries: Elucidation of Ionic Transport and Intercalation Reaction Limitations Via Electrochemical Analyses

Recently, novel rechargeable aluminum-graphite batteries using $AlCl_3$:[EMIm][Cl] ionic liquid electrolytes have been demonstrated, which store charge when chloroaluminate anions (e.g., $AlCl_4^-$) electrochemically intercalate into graphitic electrodes. However, technological development of this emerging battery chemistry is limited by incomplete scientific understanding of its charge storage mechanism, particularly since the intercalation of molecular chloroaluminate anions differs greatly compared to the intercalation of atomic cations (e.g., Li^+, Na^+, Mg^{2+}, etc.). While graphite electrodes have worked very well as an intercalation host for monatomic Li
graphite batteries, which involve intercalation of sterically-bulky, $AlCl_4^-$ molecular anions with ionic diameters (~5.28 Å) that are larger than the interlayer spacing between adjacent graphene layers (~3.35 Å).

Here, we study the transport behavior of chloroaluminate ions into graphite by comparing the electrochemical behavior at various cycling conditions of aluminum batteries containing natural graphite (NG), synthetic graphite (SG) and pyrolytic graphite (PG) electrodes. Variable-rate cyclic voltammetry (CV) revealed that the electrochemical reaction is controlled by a mixture of diffusion and other processes (e.g., kinetic or non-Nernstian). Analysis of how the CV peak current scales with varying CV scan rate (i.e., $i_p \sim v^b$) suggests that the intercalation processes exhibit different extents of diffusion-limitation at different cell voltages. These potential-dependent transport regimes vary with graphite type and are correlated with surface area and population of edge sites. Our measurements revealed a remarkable departure from lithium cation intercalation into graphite, which have been shown to exhibit a classical diffusion-limited scaling over similar scan rates. This analysis enables a quantitative comparison of how mass transport affects the rate of electrochemical intercalation of chloroaluminate species among the three graphite types, which correlates with their performance in rechargeable aluminum-graphite batteries.

566808 Modeling the Transport of Nitrogen in Gas Diffusion Electrode (GDE) for the Electrochemical Ammonia Synthesis at Ambient Conditions

Ammonia is as an effective hydrogen storage medium as it has higher hydrogen storage capacity (17.65 %) when compared to water (11.11%) and it can be transported easily for mobile applications. Ammonia can be electrochemically oxidized to hydrogen using renewable energy whenever required. So, the excess renewable energy can be used to synthesize ammonia electrochemically, apart from storing in batteries.

Ammonia is synthesized electrochemically from nitrogen and water in an alkaline medium in the presence of an electro catalyst. The challenge in electrochemical synthesis of ammonia is the low Faradaic efficiency of ammonia, which can be overcome by efficient design of electrochemical cell and an appropriate electro catalyst. The nitrogen, water and electrocatalyst must meet at the interface for the reaction to occur. Typically, planar electrode is employed by placing it in the alkaline medium saturated with nitrogen. This configuration suffers from low efficiency, as the solubility of nitrogen in water is extremely small and it is severely limited by the mass transport resistance across the boundary layer. Alternatively, gas diffusion electrode (GDE) can be used to overcome this issue. The catalyst is electrodeposited on a gas diffusion layer. Nitrogen and electrolyte are introduced on the opposite sides of the GDE. Only gas diffuses through the electrode and the electrolyte diffusion is prevented by using a hydrophobic coating on the GDE. Since hydrodynamics plays an important role in such a system, it is important to study its effect on electrochemical ammonia synthesis

In this work, a multi-physics model of nitrogen transport through a GDE is developed and solved in COMSOL. The model encompasses Nernst-Planck equation to capture the species transport in the liquid electrolyte, Darcy's law for the flow of gas in a porous medium and Tafel equations to predict the charge transfer kinetics at high over potentials. The cell performance is dependent on several parameters such as the flow rates of the gas and the liquid, orientation of the GDE, catalyst loading, thickness of the catalyst layer deposited and porosity of the catalyst. The interplay between the reaction kinetics and the species transport is studied and the operating conditions are determined to optimize the performance of the electrochemical cell. The simulations are used to study the inherent trade-offs between mass transport and reaction kinetics and to design an effective electrocatalyst-electrochemical cell system.

571734 Hexachloroiridate As an Alternative to Hexacyanoferrate in Electrochemical Mass Transfer Measurements

An important characteristic in the design of electrochemical reactors is the mass transfer performance. By using what is known as the limiting current density method, it is possible to determine the mass transfer performance of a given reactor experimentally [1]. In many cases, the hexacyanoferrate(III)/hexacyanoferrate(II) redox pair is used for this procedure. However, it is known that this couple suffers from non-ideal behavior [2-4]. Over time, the hexacyanoferrate complex will deposit on the electrodes, forming an additional resistance to mass transfer. In order to obtain reliable measurements, it is therefore necessary to periodically polish the electrodes. The situation is especially dire for electrode materials such as stainless steel, where polishing is required every two hours [5]. For 3D electrodes this is troublesome, as their fine structures cannot be polished reliably. Extensive strategies are described in literature in order to alleviate the problem, mostly with regards to the experimental conditions [6]. Despite these, we experience that the non-ideal behavior persists (see figure 1). For this reason we propose the use of the hexachloroiridate(IV)/hexachloroiridate(III) redox pair as an alternative.

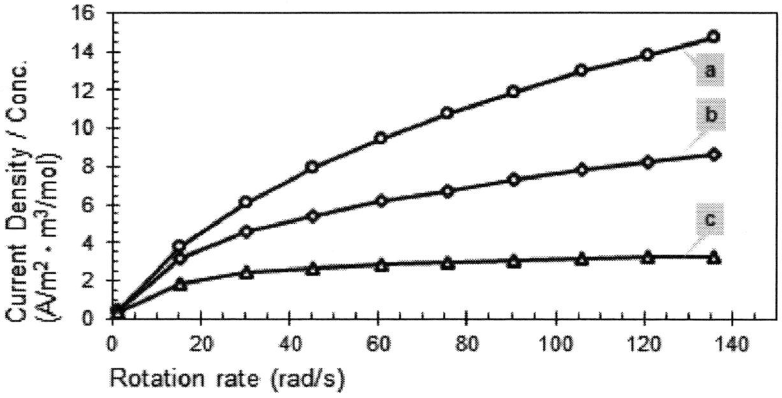

Figure 1: Limiting current density of hexacyanoferrate(III) as observed in a nickel rotor-stator spinning disk electrolyzer. a) at the start of the experimental work, b) after a day of work, c) after two days. Fresh solutions were used each time, with the recommendations of D.A. Szánto et al. being followed as far as practicable [6].

In this work we show that hexachloroiridate(IV)/hexachloroiridate(III) is a well-behaved redox couple that can be used in mass transfer experiments (see figure 2). Furthermore, its behavior is compared to that of the hexacyanoferrate couple. In contrast to the traditionally used hexacyanoferrate, we see that hexachloroiridate does not poison the electrodes (figure 3). With regards to the experimental conditions, some of the pitfalls and downsides are explored in using this new couple. Here it is shown that the pH of the solution is vitally important to the performance of the hexachloroiridate couple, which is not the case for its traditionally used counterpart. Though, through proper care, we show that the limiting current density can remain stable for over 16 hours as long as the bulk concentration is corrected for.

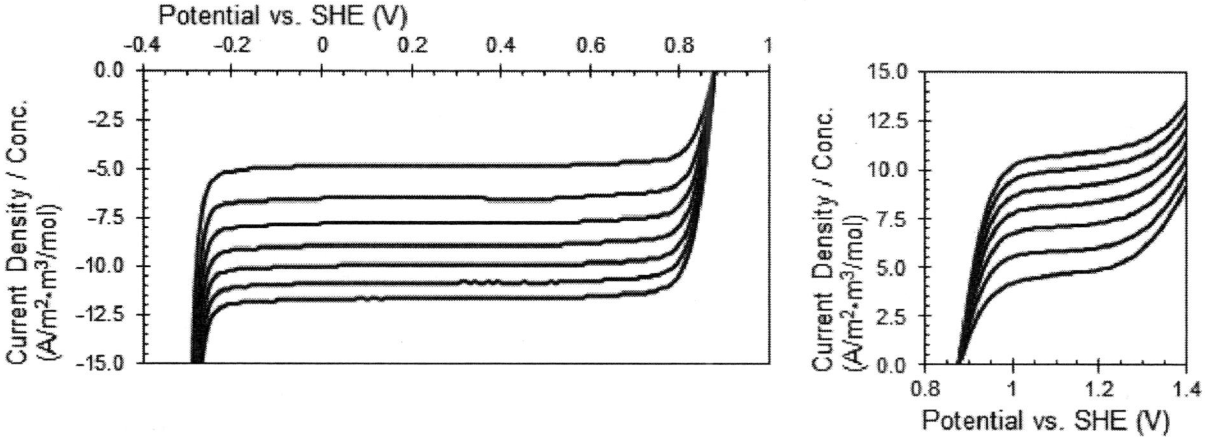

Figure 2: The limiting current density plateau of hexachloroiridate on platinum. On the left: the reduction of hexachloroiridate(IV), on the right: the oxidation of hexachloroiridate(III). A flat current density plateau is found between -0.2V and 0.7V for the reduction, and 1.0V and 1.2V for the oxidation.

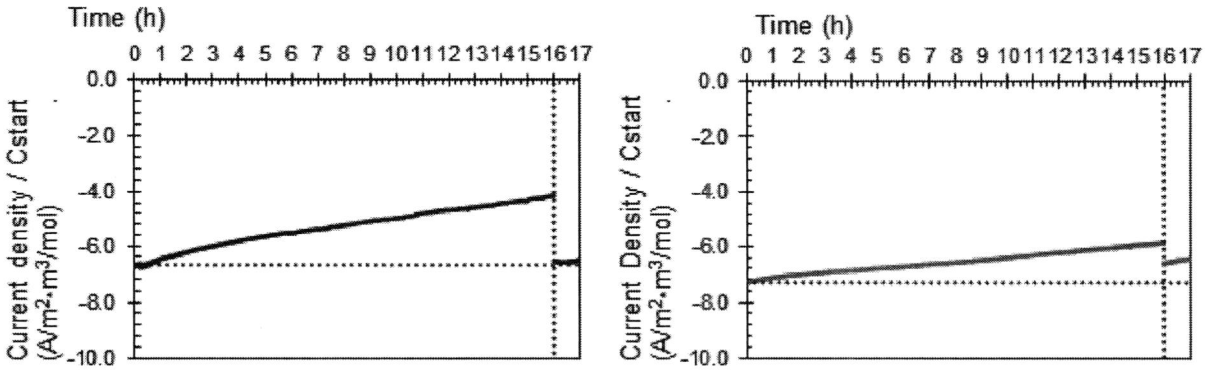

Figure 3: Evolution of the limiting current over time on platinum. On the left: the hexachloroiridate(IV) reduction, on the right: the hexacyanoferrate(III) reduction. At 16 hours, the solution was refreshed. Though the hexachloroiridate complex loses current faster, it recovers fully after refreshing. The current loss is entirely due to a loss of bulk concentration. The hexacyanoferrate current does not recover fully, which indicates some electrode poisoning.

References

[1] T. Mizushina, *"The Electrochemical Method in Transport Phenomena,"* Advances in Heat Transfer, vol. 7, pp. 87-161, 1971

[2] C. Beriet and D.J. Pletcher, *"A Microelectrode Study of the Mechanism and Kinetics of the Ferro-/Ferricyanide Couple in Aqueous Media,"* J. of Electroanal. Chem., vol. 361, pp. 93-101, 1993.

[3] M. Stiebl and K. Jüttner, "Surface Blocking in the Redox System Pt/[Fe(CN)6]3-,[Fe(CN)6]4-," Journal of Electroanalytical Chemistry, vol. 290, pp. 163-180, 1990.

[4] J. Kawiak, P.J. Kulesza and Z. Galus, *"A search for conditions permitting model behavior of the [Fe(CN)6]3-/4-system,"* J. Electroanal. Chem., vol. 226, pp. 305-314, 1987.

[5] W. Taama, R. Plimpley and K. Scott, *"Influence of Supporting Electrolyte on Ferricyanide Reduction at a Rotating Disc Electrode,"* Electrochimica Acta, vol. 41, no. 4, pp. 549-551, 1996.

[6] D.A. Szánto, S. Cleghorn, C. Ponce-de-Léon and F.C. Walsh, *"The Limiting Current for Reduction of the Ferricyanide Ion at Nickel,"* AIChE J., vol. 54, no. 3, pp. 802-810, 2008.

573157 Ptir/IrO_x on Fluffy TiO_2 /CNTs As a Catalyst for Methanol Oxidation

A new strategy for Iridium decoration with Pt nanoparticles by loading them onto a fluffy 10-nm titanium dioxide (TiO_2) surface via aqueous solution reduction method with 20 wt.% of Pt and Ir combined. As a prepared catalyst, denoted PtIr/ IrO_x-C-TiO_2/CNTs, was used for methanol electro-oxidation (MOR). The purpose is to load those metals onto a stable, conductive, and anticorrosion support of C-TiO_2/CNTs. It was found that in this study the TiO_2 coating plays a crucial role by offering a high geometrical surface area that makes optimum support to host the metal nanoparticles with small nanoparticle size. Having those metals deposited onto doped TiO_2 with small size particles has increased the active site population toward methanol molecules and enhanced the catalyst activity. Based on electrochemical results of MOR, PtIr/IrO_x-C-TiO_2/CNTs has shown a significant improvement in the anodic to cathodic current ratio with a 5.6-fold and a negative shift in onset by 120 mV relative to E-TEK (Pt/C). Furthermore, both metals shared electrons in back donation with the oxide support as a positive shift in their binding energy as revealed by XPS. We may attribute the high forward/backward current ratio and high current after one hour to hydrated iridium oxide phase incorporated on reduced Ti_xO_{2-x} where hydroxyl ligand is formed supplying protons to Pt-CO_{ads}. All other characterizations and the analyses such as X-ray diffraction (XRD), transmission electron microscopy (TEM), electron spectrum (EELS) were carried out to further understand the prepared catalyst.

Critique of Maximum Power-Point Tracking in Open-Cathode Direct Methanol Fuel Cells

Zuhair Alyousef[1] and Oscar D. Crisalle[2]

Department of Chemical Engineering, University of Florida
Gainesville, FL 32611, United States

[1]zyousef777@ufl.edu, [2]crisalle@che.ufl.edu

2019 *AIChE Annual Meeting*

Direct methanol fuel cells (DMFCs) are potential candidates for powering portable electronic devices because they are characterized by high energy densities, low operational temperatures, and because they can be physically realized in a compact design that is easy to transport by an individual user [3]. The produced electrical current can be used to power portable electronics such as batteries and it is critical to manage and optimize the delivered power to those devices. A control technique gaining emerging interest in the literature seeks to dynamically adjust the power set-point to make it match the maximum-possible power output of the device at the current operating temperature. These maximum-power-point tracking (MPPT) algorithms have enjoyed significant success in photovoltaic (PV) power-generation applications, and looks appealing for fuel cells at first glance. The purpose of this study is to analyze the advantages and disadvantages of MPPT and advance the argument to compare it with a proposed alternative of maximum efficiency point tracking (MEPT) strategy.

Figure 1 shows a schematic diagram of the DMFC system considered in this work. The system follows an open-cathode architectural design, where a fraction of the product water is vented to the atmosphere. The core components are a mixing tank containing a diluted methanol solution that serves as the fuel. The fuel-cell stack is represented in the figure by the area located between the anode flow channel (AFC) that transports the liquid fuel, and the cathode flow channel (CFC) transporting the air. A prominent feature of the stack is the polymer exchange membrane (PEM) serving as a porous medium for water and ions. A fuel cartridge provides a pure-methanol stream available on demand, and a fan is used to draw air through the cathode flow channel for the dual purpose of supplying oxygen to the stack and adjusting the operating temperature of the device.

Power control in a DMFC is a critical element needed for a successful operation. The control strategy of MPPT is well established in PV cells, where it has proven to increase the solar-to-electric conversion efficiency under real-time implementation [2]. Following this success in PV systems, several studies have been done to implement MPPT algorithms in polymer electrolyte membrane fuel cells and DMFCs [4, 5, 6]. While operating at the maximum-power point in a DMFC is clearly advantageous in terms of reducing the extent of methanol crossover and improving fuel efficiency, further investigation is required to determine if there still a net benefit when other fuel-cell performance metrics are taken into consideration. Examples of such criteria include the analysis of likelihood of cathode flooding under MPPT and the effect of the algorithm on the overall device efficiency. This study suggests that the alternative control strategy, namely MEPT is a superior one.

Figure 2 shows characteristic power-polarization curves for a DMFC. The vertical axis represents the power P in watts [W] while the horizontal axis represents the current I drawn from the cells in units of ampere [A]. Each one of the curves is a power isotherm, *i.e.*, the power produced at a constant temperature. At each polarization curve, there is only one maximum-power point (MPP) at the peak indicated by a

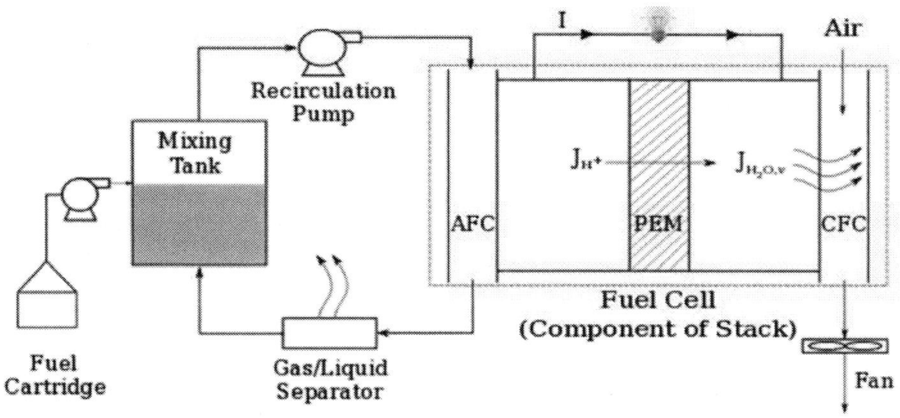

Figure 1: Schematic diagram of the DMFC system considered for this study, where a mixing tank filled with a methanol fuel solution is piped thorugh a circulation loop with a fuel cell stack that produces electrical power by generating an electrical current I.

filled circle markers. It should be emphasized that those curves are produced by varying the temperature of operation, while keeping constant the mixing tank's methanol concentration and the fuel flow rate to AFC.

As a demonstration of the maximum-power-point tracking principle, suppose that the power required by the external load is 43 W. There are many feasible options for operation at this power level, among which are points A and B, which respectively denote the power points for the polarization curves at 50 and 55 °C. However, only point A is a maximum-power point, for the load requirement in question. The task of the MPPT control algorithm is to search for this MPP and adjust appropriate variables, namely the current drawn and the fuel-stack temperature, to keep the cell operating at that point in a persistent fashion.

Several advantages are realized by operating the fuel cells at the maximum-power point for a given load demand. For demonstration purposes, consider points A and B. First, while both points A and B deliver the same output power, MPP A operates at a lower temperature of 50 °C, which leads to less methanol crossover [1]. Second, the current drawn from the cell at A is higher than that drawn at B by approximately 20 %. At point A, the cell produces higher *fuel efficiency* defined as

$$\varepsilon_{fuel} = \frac{I}{I + I_x} \tag{1}$$

where I_x is the methanol crossover current. Furthermore, the implementation of a real-time tracking of MPP is feasible using standard instrumentation hardware since both current and cell voltage E are measurable.

Despite those remarkable benefits of MPPT, several concerns are of significance. A first concern is that, as is evident from inspection of the power curves in Figure 1, the current corresponding to the MPP A is very close to the limiting current of 3.8 A, defined as the zero-power point on the curve where the device stops functioning. Therefore, operation at the MPP risks producing power failures when the system is subjected to perturbations that temporarily move the operation to currents located to the right of the MPP current. A second concern is that the close proximity of the current at the MPP to the limiting current implies that the operating regime is inside the mass-transport-limited region of the current-voltage polarization curve, where significant voltage losses are observed for small increases in the current. Hence, when operating at the MPP A produces a lower the *voltage efficiency*

$$\varepsilon_{voltage} = \frac{E}{E_{thermo}} \tag{2}$$

where E is the actual voltage of the cell and E_{thermo} is the thermodynamic voltage of the cell that defines the maximum theoretically-possible voltage. A third problem of concern is that water is produced at a higher rate by the chemical reaction **??** when the electrical current (represented by the rate of electron generation supplied as reagents on the left-hand side of the equation) is high. Hence it is likely that flooding occurs in the pores of the membranes in the cathode region of the fuel cells in the stack, completely blocking the

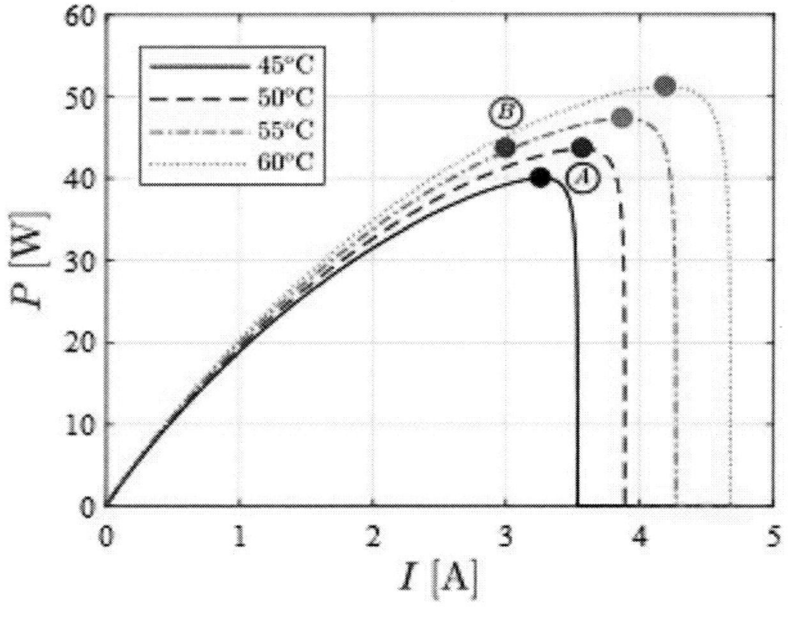

Figure 2: Power diagram in DMFC, with current I as x-axis and power P as y-axis.

transport of oxygen supplied as a reagent to chemical reaction ??. The 100 % liquid saturation limit is a dangerous operating regime because it causes oxygen starvation in the cathode, leading to a failure in the form of an abrupt halt of device operation.

Figure 3 shows the curves for the overall efficiency ε of the DMFC, where

$$\varepsilon = \varepsilon_{fuel}\varepsilon_{voltage}\varepsilon_{thermo} \tag{3}$$

where $\varepsilon_{thermo} = \frac{\Delta G}{\Delta H}$ is the ratio of the change of Gibbs energy to the change in enthalpy of the overall electrochemical reaction. The vertical axis represents the efficiency ε in % while the horizontal axis shows the current I in Amperes. Each one of the curves is an efficiency isotherm. The maximum-efficiency point (MEP) for each isotherm is indicated by a circle marker. Operating at MEP mitigates or altogether eradicates the concerns applicable to the MPPT algorithm. First, the issue of proximity to the limiting current is no longer a relevant issue in this case since the MEP points are at located at least 1.2 A away. Additional simulation results (not shown here for brevity) demonstrate that under MEP operation the saturation of the pores with liquid water ranges between 40 and 60 %, indicating that cathode-side flooding is unlikely.

References

[1] M.A.R. Biswas, S.P. Mudiraj, W.E. Lear, and O.D. Crisalle. Systematic approach for modeling methanol mass transport on the anode side of direct methanol fuel cells. *International Journal of Hydrogen Energy*, 2014.

[2] Lijun Qin and Xiao Lu. Matlab/simulink-based research on maximum power point tracking of photovoltaic generation. *Physics Procedia*, 24:10–18, 2012.

[3] F. Zenith and U. Krewer. Modelling, dynamics and control of a portable dmfc system. *Journal of Process Control*, 20:630–642, 2010.

[4] Mingbo Zhang, Ting Yan, and Jinguang Gu. Maximum power point tracking control of direct methanol fuel cells. *Journal of Power Sources*, 247:1005–1010, 2014.

[5] Zhong Zhi-dan, Huo Hai-bo, Zhu Xin-jian, Cao Guang-yi, and Ren Yuan. Adaptive maximum power point tracking control of fuel cell power plants. *Journal of Power Sources*, 176:259–269, 2008.

[6] Guo-Rong Zhu, KH Loo, YM Lai, and K Tse Chi. Quasi-maximum efficiency point tracking for direct methanol fuel cell in dmfc/supercapacitor hybrid energy system. *IEEE Transactions on Energy Conversion*, 27(3):561–571, 2012.

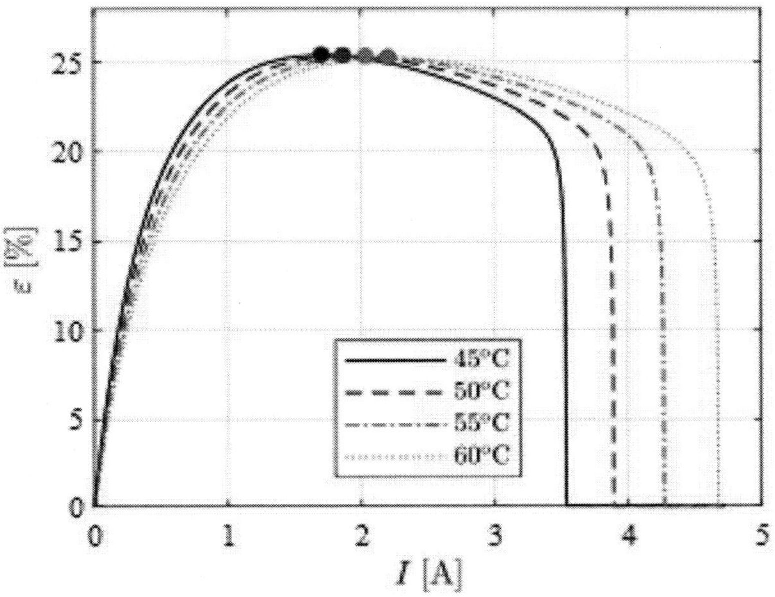

Figure 3: Overall efficiency diagram in DMFC, with current I as x-axis and efficiency ε as y-axis.

576132 Investigation of the Microstructure and Rheology of Iridium Oxide Catalyst Inks for Low-Temperature Proton Exchange Membrane Water Electrolyzers

Low-temperature polymer electrolyte membrane water electrolyzers (PEMWE) are an attractive clean energy technology to produce hydrogen (H2) which is an energy carrier for several applications such as transportation and grid-scale energy storage and distribution (as supported by the US Department of Energy's H2@Scale initiative). A critical component of PEMWE membrane electrode assemblies (MEA) is the catalyst layer (CL), where the electrochemical reactions occur. The microstructure of the CL plays a key role in MEA performance by affecting catalyst utilization, proton conductivity and gas transport. The CL microstructure is formed by the catalyst particles and an ionomer - which acts both as a binder for catalyst particles and as a proton conducting medium. The evolution of CL microstructure is primarily controlled by the nature of fabrication of CL, which is commonly solution processing of an ink dispersion. In this process, an ink is formulated at a specified catalyst and ionomer ratio, dispersed in water-alcohol solvent, and then coated (via. spray-coating, roll-coating or brush-painting) onto a substrate (polymer electrolyte membrane or decal) and dried to form the CL. The interactions between the ink components (catalyst-ionomer-solvent), and the ionomer concentration in the ink, which dictate the ink microstructure and the processing behavior during coating and drying, play a critical role in the evolution of final CL microstructure. In this talk, an investigation of catalyst-ionomer interactions, and the influence of alcohol-water solvent mixture on microstructure and rheological behavior of the catalyst inks will be presented. The ink consists of iridium oxide (IrO_2) catalyst particles and Nafion ionomer dispersed in a mixture of 1-propanol and water. A combination of rheology, dynamic light scattering, electrophoretic mobility and ultra small-angle X-ray scattering (USAXS) techniques will be used to understand inter-particle/particle-ionomer interactions and characterize microstructure. Our initial findings suggest that iridium catalyst dispersions in the absence of ionomer are agglomerated. The addition of ionomer stabilizes the particles through adsorption via the electrosteric mechanism. The stability and microstructure evolution as a function of ionomer concentration in the ink will be discussed.

571759 Demystifying Mass Transfer in Electrolyzers through 3D Printed Parallel Plate Cells: The Importance of Inlet Effects

As a result of the global energy transition, a large percentage of the world's power supply will become electric. For electrochemical engineers this transition generates a sea of opportunities. After all, it is electrolyzers that use electricity to produce some of life's essential chemicals. However, fully embracing these opportunities in a sustainable way is challenging: Future electrolyzers must be safer, cleaner, and far more flexible, while simultaneously increasing productivity and cost-effectiveness. Essentially, they will have to achieve more with less.

One key parameter in the performance of electrolyzers is the rate of mass transfer. Most often, electrochemical reactions are mass transfer limited due to the fast kinetics that depend exponentially on the applied overpotential. Numerous correlations in literature exist to describe the rate of mass transfer in a specific reactor.[1] Though, their predictions can vary up to almost an order of magnitude (see figure 1). The reason for this is the large geometric variety between the reactors that were measured. The design of the inlets in particular is a crucial contributor to the rate of mass transfer. [2] Moreover, only a few lab-scale electrolyzers are commercially available, often with sub-optimal designs, which limits further research. As a consequence, it is difficult to establish a generalized description of the mass transfer performance of parallel plate electrolyzers. Furthermore, most existing correlations only describe the average rates of mass transfer and therefore do not consider any local variation despite the importance thereof.[3]

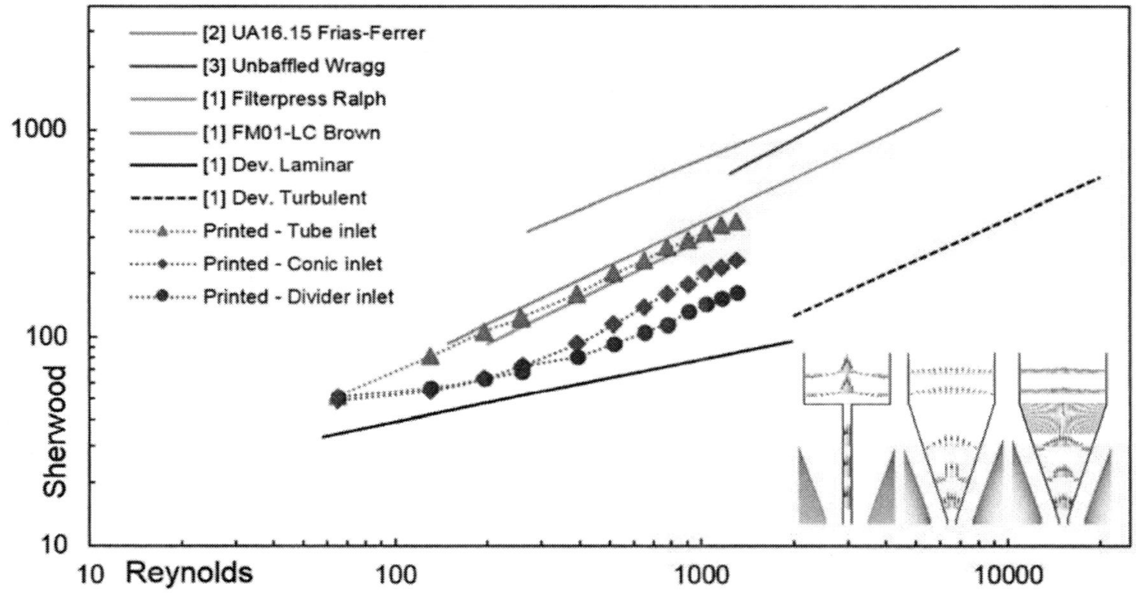

Figure 1: Mass transfer performance in terms of Sherwood vs. Reynolds. Up to an order of magnitude of difference in performance is visible between reactors. The red triangle, square and circle markers denote a selection of inlet manifolds tested with the 3D printed prototype. Small insert: sketches of the inlet manifolds, from left to right: tube inlet, conic inlet, divider-type inlet.

The goal of our study has been to demystify the fundamentals of mass transfer in parallel plate electrolyzers. As this would have required testing a large geometric variety of cells, traditional manufacturing techniques were impractical due to their lead time and cost. Therefore, we have used 3D printing to rapidly construct many different reactor prototypes. Moreover, each prototype could be finetuned to the exact specifications each measurement required. This way, each geometric parameter could be varied and investigated individually.

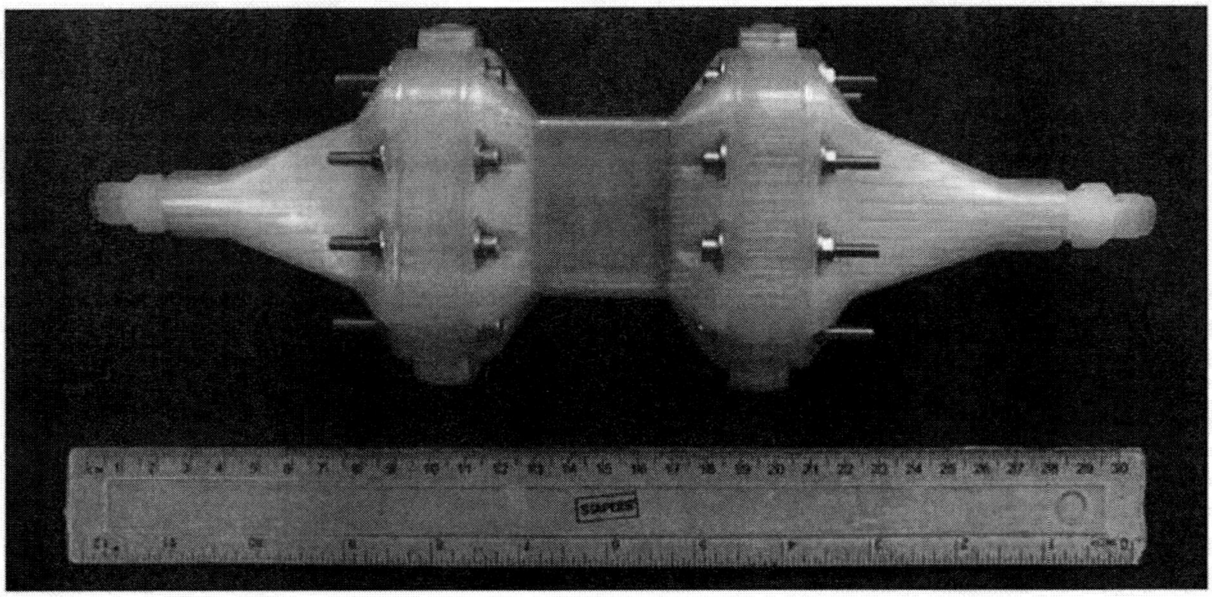

Figure 2: 3D printed electrolyzer prototype for mass transfer studies. The electrode area is 40 cm^2.

Nearly the entire reactor can be 3D printed in less than a workweek's time (see figure 2). Inlet manifolds, the channel and the rubber sealings could all be printed using commercially available equipment. Furthermore, by using 32 independently operated electrode segments we were able to measure the mass transfer performance on a local scale. Sherwood – Reynolds correlations were established for designs with different shapes of turbulence promoters, 3D electrodes and inlet manifolds. The effect of an increased or decreased hydrodynamic entrance length was also investigated.

In the end these prototypes lead to enhanced understanding of electrochemical reactors, allowing us to intensify processes such as alkaline water electrolysis. As a proof of concept, we are developing a 3D printed lab-scale electrolyzer for alkaline water splitting which can be used to test the effect of geometry on the production of hydrogen directly.

References

[1] T.R. Ralph, M.L. Hitchman, J.P. Millington, F.C. Walsh, *Mass Transport in an electrochemical laboratory filterpress reactor and its enhancement by turbulence promoters,* Electrochem. Acta, **1994,** pp. 591-603.

[2] A. Frías-Ferrer, J. González-García, V.Sáez, C. Ponce de León, F.C. Walsh, *The effects of manifold flow on mass transport in electrochemical filter-press reactors,* AIChE J., **2008,** pp. 811-823.

[3] A.A. Wragg, A.A. Leontaritis, *Local mass transfer and current distribution in baffled and unbaffled parallel plate electrochemical reactors,* Chem. Eng. J., **1997,** pp. 1-10.

565010 Ni-Based Cermet Anode with a Three-Dimensionally Ordered Macroporous Structure for Solid Oxide Fuel Cells

Solid oxide fuel cells (SOFCs) are promising energy devices with high energy efficiency and low emissions [1]. Using hydrocarbons instead of hydrogen as fuels is one of the major advantages of SOFCs. The most common anode of SOFCs consists of metallic Ni and an ionic conducting ceramic phase. However, Ni suffers from not only aggregation at a high temperature, but also serious carbon deposition with hydrocarbons as fuels [2]. The optimization of anode structure is recognized as an efficient way to solve those problems.

In this work, Ni-samarium doped ceria (SDC) with a well-controlled structure is studied as an anode material of SOFCs. SDC framework with a three-dimensionally ordered macroporous (3DOM) structure was synthesized using close-packed poly(methyl methacrylate) (PMMA) spheres as the template. 10 wt% Ni catalyst was added in the SDC scaffold through impregnation. The agglomeration of Ni nanoparticles in the anode is effectively prevented by the "confinement effect" of the SDC framework. X-Ray diffraction results and transmission electron micrographs confirm the formation of $Ce_{1-x}Ni_xO_2$ solid solution, which induces abundant oxygen vacancies. The investigation on the structure and surface properties of Ni/3DOM-SDC composite anode reveals the synergetic effect between Ni and SDC phases, facilitating the electro-oxidation of H_2 and methanol fuels, and improving the resistance to coking. A single cell with Ni/3DOM-SDC anode and SDC-carbonate composite electrolyte exhibits a maximum power density of 1633 mW cm^{-2} at 700 °C with gasified methanol as the fuel, much higher than that of the cell with conventional Ni-based anodes without the 3DOM structure under the same condition.

References

[1] Z. Zhan, S.A. Barnett, Science, 308 (2005) 844.

[2] A.A.A. da Silva, N. Bion, F. Epron et al., Appl. Catal., B, 206 (2017) 626.

579977 Highly Efficient Sorbent Systems for Carbon Capture

TDA Research, Inc (TDA) is developing a low cost, high capacity CO_2 adsorbent and demonstrating its technical and economic viability for CO_2 capture from coal-fired power plants. TDA uses an advanced physical adsorbent to selectively remove CO_2 from the flue gas. The sorbent exhibits a much higher affinity for CO_2 than N_2, H_2O or O_2. The sorbent binds CO_2 more strongly than common adsorbents, providing the chemical potential needed to remove the CO_2. However, because CO_2 does not form a true covalent bond with the surface sites, the regeneration can be carried out with only a very small energy input (4.9 kcal per mol of CO_2) compared to amines (14.2 kcal/mol CO_2). We demonstrated the sorbent's efficacy through multiple cycles for removing CO_2 from coal derived flue gas and the sorbent achieved a stable capacity. The presence of acid gases such as SO_2 and NOx (80 and 200 ppmv, respectively) and water vapor up to 15% vol. caused no adverse effect on the CO_2 capacity. The regeneration of the sorbent and the recovery of CO_2 and its pressurization can be achieved by several approaches, including temperature and vacuum swing. A detailed techno-economic analysis using Aspen Plus^{TM} software is conducted in collaboration with University of California Irvine to estimate the impact of the TDA's CO_2 capture system on plant efficiency and cost of electricity. Our system analysis showed that TDA's CO_2 capture process will reduce plant efficiency only by 19.4% (compared to 31.2% reduction for amine scrubbers) while providing 90% capture and the increase in 1st year levelized cost of electricity (COE) is 18.4% point lower than that of amines. We will present the latest results from the DOE sponsored work at the AIChE meeting.

556157 Transformational Graphene Oxide-Based Membrane Process for Post-Combustion CO_2 Capture

The Department of Energy (DOE), through its Carbon Capture Program, is investigating transformational, low-cost CO_2 separation and capture technologies that can be installed in new or retrofitted into existing pulverized coal (PC) power plants for CO_2 capture with a cost of electricity at least 30% lower than a supercritical PC with CO_2 capture, or approximately $30 per tonne of CO_2 captured.

We are developing a transformational graphene oxide (GO)-based membrane process that can be installed in new or retrofitted into existing pulverized coal (PC) or natural gas power plants for CO_2 capture with 95% CO_2 purity. The GO-based membrane technology builds on our pioneering work on the ultrathin, GO-based membranes published in *Science* (Li et al., *Science*, Vol. 342, no. 6154 (2013) 95-98) and *Nature Communications* (Zhou et al., *Nature Communications*, Vol. *8,* doi:10.1038/s41467-017-02318-1342).

A facile and scalable coating process was developed to prepare ultrathin graphene oxide (GO)-based membranes on polyethersulfone hollow fibers for highly efficient CO_2/N_2 separation. The GO-based membranes developed to date exhibited extraordinary separation performance under simulated flue gas conditions with CO_2 permeance of 1,020 GPU and a CO_2/N_2 selectivity as high as 680. The selectivity is much higher than state-of-the-art membranes. In addition, after extensive material and structure optimization, we have also developed high-flux membranes with CO_2 permeance as high as 2,500 GPU using GO quantum dots (GOQD) as a membrane building block.

Our transformational GO-based membrane process (designated as GO^2) integrates the high-selectivity and high-flux GO-based membranes and offers a new opportunity to explore further reductions in the cost of CO_2 capture. The GO^2 process, functionally, has the benefits of a hybrid system but the simplicity of a membrane system which also reduces the up-front installation costs and footprint and does not add circulating liquids to the power plant environment. Therefore, it facilitates the ease of integration into a power plant.

576662 Water Lean Amine-Based Solvents for Carbon Dioxide Capture

Carbon dioxide released from large point sources such as coal powered plants should to be captured, stored or utilized for environmental sustainability. In order to achieve this goal, many CO_2 capture technologies have been developed, with the most advanced being aqueous amine-based solvent process. These aqueous amine solvents have low CO_2-rich viscosity, fast kinetics, and favorable thermodynamics. Deployment of these aqueous amine solvents to market is limited by high regeneration temperature due to the high heat capacity of water. To address this limitation our group and others have developed water-lean solvent systems. PNNL has developed water-lean amine-based solvents utilizing key fundamental learnings from multiple solvent chemistries. These amine-based solvents have high gravimetric CO_2 capacity (12-15 wt%) and form liquid carbamates with significantly lower CO_2-rich solvent viscosities (~100 cP) and were regenerated at low temperatures (80-100 °C). These solvents can be recycled over multiple CO_2 absorption and regeneration cycles without significant evaporative loss or degradation. Herein we will discuss the molecular design, synthesis, characterization and testing of these amine solvents.

571716 CO_2 Absorption in Amines Using SiO_2 Nanoparticles in Instantaneous Regime

Nowadays, increasing levels of CO_2 in atmosphere is serious concern worldwide. This is caused due to the imbalance between the rate of emission of CO_2 and its removal. Several methods have been used to reduce the effect of these emissions, major technology being CO_2 absorption in amines. Several studies carried out earlier show that, with minute changes to the present absorptive systems, we can increase the absorption rates with would also help us in design of equipments. We have, therefore explored the application of principle of mass transfer enhancement by nanoparticles to the problem of absorptive capture of CO_2 in reactive systems.

From the literature, it is seen that concentration of solvents as well as the concentration of nanoparticles both affect the absorption rate. With optimum concentrations of both, we can achieve a state of maximum possible enhancement. For this, we have studied the effect of nanoparticles on CO_2 absorption in MEA studied in a packed column reactor in presence of Silica nanoparticles which is of interest from an industrial application point of view.

Mono Ethanol Amine (MEA) is chosen as a solvent and the parameters are adjusted to fit in to the Fast to Instantaneous regime and Instantaneous regime. Experiments are performed at different volume fractions of SiO_2 nanoparticles and concentrations of MEA. Similar experiments are conducted in capillary tube model apparatuses to check the consistency of mechanisms between both the systems.

It is observed that fewer enhancements are obtained in Instantaneous regime as compared to the Fast to Instantaneous regime. The highest possible enhancement has been found at a SiO_2 particle volume fraction of 0.6 above which agglomeration of nanoparticles is observed. The enhancement due to nanoparticles has been evaluated using some of the existing models from the literature and deviations from the experimental ones are reported. A new model has been proposed which takes in to account all the parameters in this regime and can estimate the enhancements caused due to them.

576469 Plant-Wide Optimization of a Moving Bed Process for CO_2 Capture Using a Micro-Encapsulated Solvent

Micro-encapsulation of solvents is an excellent option for the potential solvents that are highly viscous such as the ionic liquids or those that form solids upon CO_2 capture/release. These solvents are operated at low CO_2 loading or under high dilution to avoid solids precipitation and for facilitating transportation between the absorber/regenerator thus significantly restricting the high potential of this large class of potential solvents. In the micro-encapsulation technology, a solvent is encapsulated in a polymer through microfluidic techniques [1]. Therefore even highly corrosive solvents can be utilized. Furthermore, micro-encapsulated carbon sorbents (MECS) also offer large surface area that improves the mass transfer and heat transfer characteristics.

For maximizing the economics of the MECS technology, design and operating conditions of an optimal contactor technology must be optimized. One of the potential technology is the conventional fixed bed technology. However, it has several disadvantages mainly due to large reactor volumes, high pressure drop, difficulty in heat recovery, and complex cyclic nature of fixed bed operation that makes heat recovery a challenge. The moving bed (MB) technology has been reported to have strong potential for solids-sorbent based CO_2 capture [2-4]. The near-countercurrent contact between the solids and gas results in an overall large driving force leading to higher CO_2 loadings from the adsorber and lower CO_2 loadings from the regenerator. Heat recovery in the MB processes is simpler compared to the fixed bed processes since the heat can be readily recovered from the hot lean sorbents leaving the regenerator using a working fluid, which then transfers the heat to the incoming loaded sorbent to the regenerator. While there are few works for the fixed bed [5-6] systems for the MECS, to the best of our knowledge, there is no published work in the open literature on the MB models for CO_2 capture using MECS.

In this work, a detailed model of a moving bed reactor containing these microcapsules with sodium carbonate as the solvent is developed. The multi-scale model couples the capsule model with the contactor-scale model using appropriate volume-averaged transport variables. The microcapsule model incorporates detailed reactions, water flux, species conservation, and energy conservation along with rigorous thermodynamic and transport properties models. Water transport through the polymer shell is also accounted for. Mass transfer parameters at the capsule level are optimally estimated using the dynamic experimental data obtained under varying temperature and pressure conditions. The contactor-scale model includes the mass, momentum and energy balances along with rigorous hydrodynamic models as appropriate for the MB operation. For heat removal in the adsorber, an embedded cooler is considered while an embedded heater is considered in the regenerator. Balance of the unit includes heat recovery heat exchangers as well as other equipment items such as the coolers and pumps for the heat transfer medium, etc. The plant-wide model is used for optimal design of the equipment items as well as their operating conditions for minimizing the equivalent annual operating cost (EAOC). It was observed that there can be significant variation in the economics of the MB process in comparison to the fixed bed process making them a great candidate for MECS-based CO_2 capture especially when a high extent of recovery of the sensible heat is desired for improving the process economics.

References

1. Vericella, J. J. et al. "Encapsulated liquid sorbents for carbon dioxide capture", Nature Communications, 6:6124, 1-7, 2015.
2. Knaebel, K.S., Temperature swing adsorption system. 2009, Google Patents.
3. Mondino, Giorgia et al. "Effect of Gas Recycling on the Performance of a Moving Bed Temperature-Swing (MBTSA) Process for CO_2 Capture in a Coal Fired Power Plant Context." Energies, 10(6), 745. 2017.
4. Kim et al. "Moving bed adsorption process with internal heat integration for carbon dioxide capture." IJGGC, 13-24, 2013.

5. Anggit Raksajati, et al, "Techno-economic Evaluation of CO_2 Capture from Flue Gases Using Encapsulated Solvent", Industrial & Engineering Chemistry Research 1604-1620, 56 (6), 2017.
6. Hornbostel, et al, Packed and fluidized bed absorber modeling for carbon capture with micro-encapsulated sodium carbonate solution, Applied Energy 1192-1204, 2019.

569614 CO_2 Utilization Via Chemical Looping Dry Reforming: Process Modeling and Systems Analysis

CO_2 capture and storage or sequestration (CCS) is one of the key climate change mitigation technologies. In CCS, CO_2 is captured from flue gases of power plants or industrial sources, and the high-purity CO_2 is compressed and stored in deep geologic formations. However, this process is associated with high energy consumption and cost. Utilizing the captured CO_2 instead for conversion into valuable chemicals and fuels could not only (partially) close the anthropogenic carbon cycle, but also be a cost-effective alternative to make CCS technology commercially viable. Among CO_2 utilization routes, dry reforming of methane to syngas is promising since it couples CO_2 utilization to fuels production, one of the few markets with a scale to absorb the huge amounts of CO_2 emitted from power generation. We have previously proposed "chemical looping dry reforming" (CLDR) as a promising process alternative to conventional dry reforming [1-2], and later extended this approach towards an alternate process configuration that directly produces fully separated syngas streams, i.e. separate effluent streams of H_2 and CO [3-4]. Like all "looping" processes, CLDR is based on the use of spatially or temporally separated reactors for fuel conversion with an oxide carrier followed by re-oxidation of the carrier with an oxidant (here: CO_2). In the present paper, we analyze the feasibility of CLDR by combining the dry reforming reaction with power generation using waste heat recovery.

In our study, the overall system is designed to produce a desired flow rate and composition of the syngas (i.e. the CO/H_2 ratio) from natural gas, while maximizing the yield and minimizing the overall energy consumption. The CO_2 feed is assumed to come from a CO_2 capture at a power plant or other industrial process. Building on our previous fixed-bed experimental results, reactor-level mass and energy balance calculations are performed to calculate the effect of operating conditions such as flow rates, solids conversion and temperature on the final product composition, selectivity and yields. The systems-level model includes thermodynamic calculations of individual process components such as blowers, compressors, and heat recovery equipment, as well as steam turbines. Using sensitivity analysis on parameters such as selectivity and fuel conversion, we identify the carbon carrier properties required to improve process performance, and compare results to a conventional CO_2 dry reforming process as well as to conventional H_2 production pathways from CH_4 using autothermal reforming or steam methane reforming. Preliminary cost assessments will also be conducted to evaluate the comparative economic feasibility of the CLDR process.

[1] M. Najera, R. Solunke, T. Gardner, and G. Veser, "Carbon capture and utilization via chemical looping dry reforming", Chem. Eng. Res. & Des. 89 (2011) 1533-1543

[2] S. Bhavsar, M. Najera, and G. Veser, "Chemical Looping Dry Reforming as Novel, Intensified Process for CO2 Activation", Chem. Eng. Technol. 35 (2012) 1281–1290.

[3] A. More, S. Bhavsar, and G. Veser. "CO2 Activation Via Chemical Looping Dry Reforming Of Methane", Energy Technology 4 (2016) 1147–1157.

[4] A. More, Ch. Hansen, and G. Veser, "Production of inherently separated syngas streams via chemical looping methane cracking", Catal. Today 298 (2017) 21-32.

576139 Design Development of an Integrated Reforming and Gasification Technology for Syngas Production

The production of syngas is one of the important process in the petrochemical industries from which many downstream products like ammonia, methanol and fertilizers are produced. In order to have a sustainable hydrocarbon fuel usage, the amount of carbon dioxide in the atmosphere must be reduced. Syngas production technologies can play a role in bridging the gap for the use of fossil fuels while keeping the emissions within the limits. Syngas can be produced using gaseous fuels, liquid hydrocarbons and solid feedstocks. The gaseous and liquid feedstocks utilize technologies such as steam reforming, partial oxidation, dry reforming, and auto-thermal reforming. With solid feedstocks such as coal and biomass, gasification is the main technology available for the syngas production. Recently, dry methane reforming is gaining a lot of attention as an environmentally friendly process for syngas production because it consumes two major GHGs, carbon dioxide and methane. It requires 363.6 kg of methane to consume 1 ton of CO_2 for the production of 90.9 kg of H_2 and 1.27 tons of CO.

This study is focused on the production of syngas using gasification and dry methane reforming (DMR) technologies. The two processes are compared as a standalone design for a syngas production capacity of 10,000 kmol/h with H_2 to CO ratio of 0.88. The two processes are then integrated to develop a novel design that can reduce the CO_2 emissions to a near-zero level with reasonable amount of energy requirement. In this study, Aspen Plus® has been employed as a process simulation tool.

The results reveal that the standalone DMR process consumes 5.4 times more energy compared to a gasification plant of same capacity. This means that due to the high energy demand in the DMR technology, it will be uneconomical to produce syngas while mitigating CO_2 emissions. Therefore, the integration of gasification and DMR design off-sets the negative aspects of gasification technology enabling it to utilize conventional solid fossil fuel with low emission levels. The proposed integrated design in this study requires 80.5 MW energy compared to the 265 MW energy required in the DMR design. The study performed a detailed techno-economic and CO_2 emissions analysis in order to evaluate the feasibility of the proposed design.

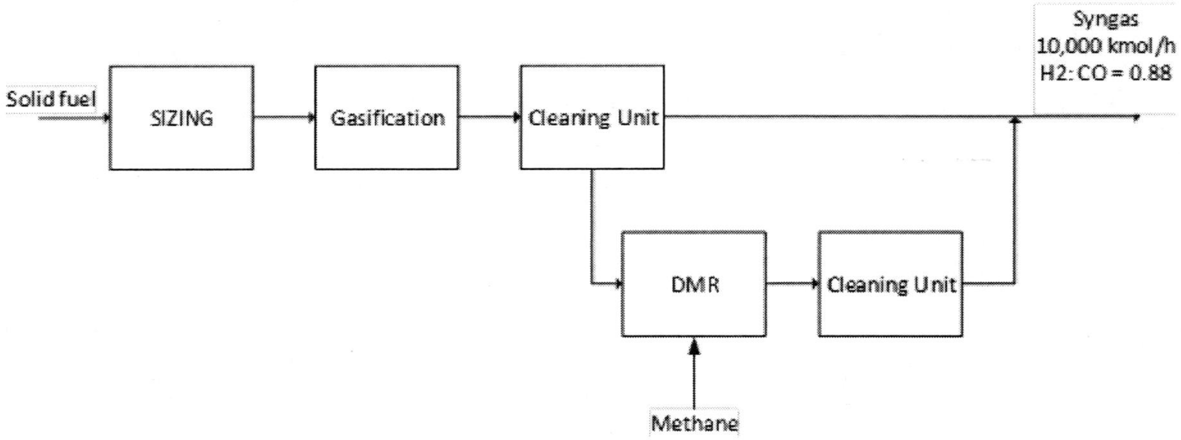

Figure 1: Integrated gasification and dry reforming design for syngas production

574517 Modal Switching for Continuous Solar Reforming

The carbon footprint of methane reforming processes can be reduced by using solar thermal energy in place of fossil energy for process heating. In this scheme, concentrated solar power provides heat that drives endothermic reformation of methane, storing renewable energy in the chemical bonds of the resultant synthesis gas. CO_2 can be directly valorized in this process to further offset end-use carbon emissions. However, solar transience hinders the operational window and overall efficiency of these systems, limiting their practical feasibility. Here predictive switching between bi-reforming and tri-reforming of methane modes over a single nickel-based catalyst is applied in order to operate solar methane reforming continuously. This is performed using Oregon State University's solar simulator and directly-irradiated volumetric reactor, which is designed to promote fast transient behavior of solar-thermochemical reactions. Solar-to-chemical efficiency is calculated over periods of transient modal switching, where solar power is reduced incrementally and oxygen with additional methane are throttled into the feed to compensate for this decrease in external heating. Catalyst stability and coke formation are examined after modal switching.

556448 Experimental Study on a Solar Falling-Film Photochemical Hybrid System for the Decolorization of Azo Dyes

Based on the solar photochemical reactors with flow channel (SPC), a new hybrid system that integrated solar falling-film photochemical reactor with photovoltaics (SFPC) was first presented to achieve the dye decolorization and the electricity generation simultaneously. Persulfate was chosen as the photochemical oxidant, all the experiments were performed outdoors in Tianjin, China. In this work, the spectrum transmittance, decolorization efficiency and the PV panel parameters were analyzed, by dealing with different concentration of acid red 26 (AR26) and acid brown 14 (AB14) in different systems.

The structural modification on the reactor improved the optical transmission efficiency of the SFPC system. The actual optical loss was only 1.6% in the SFPC system in contrast of 21.9% in the SPC system. The results confirmed that the remove of borosilicate glass and the reduced water thickness greatly decreased the optical loss in the SFPC system. The performance of the dye decolorization and the PV panel parameters were also tested. Compared with the SPC system, the SFPC system greatly enhanced the decolorization efficiency, the accumulated ultraviolet needed shortened 33.64%, the I_{sc}, P_{max} and of the PV panel increased 26.97%, 30.47%, 2.9% respectively, when dealing with the 50 mg/L AR26. And we will do the research on the degradation mechanism of the dyes in the future work.

The flow rate of the wastewater was also optimized. When the flow rate increased from 200 L/h to 1000 L/h, the reaction kinetic constants of AR26 and AB14 increased by 0.3 and 0.97 times respectively. However, the output power of the panel decreased and the energy consumption to maintain the operation of the system increased 4.93%. Taking the high energy loss with little rise of decolorization efficiency, 200 L/h is best choice for the system operation.

556434 Wax Deposition in Subsea Oil Transportation Pipelines

The gelation and deposition of waxy crude oils in subsea oil pipelines pose great threats to the safety and efficiency of their transportation. In this poster, we compile and present findings from several recent investigations regarding the gelation and deposition of waxy crude oils. We first conducted a comprehensive investigation of the rheological responses of the gelled waxy crude oils by both stress-controlled and strain-controlled tests and established a comprehensive description of the yielding scenario for waxy crude oils. We also established a numerical model to simulate the deposition rate under various conditions and used this model to elucidate the effects of two important operating conditions, namely temperature and flow rate, on wax deposition. In a recent investigation, we found that the viscosity of waxy crude oils can be reduced by more than 80% after exposed to DC electric field with only 1% of the energy assumption of conventional methods. These findings help improve the understanding of wax deposition mechniams and provide guidance for field operations.

Skills:

1. Experimental Skills: Rheology, Materials characterization

2. Modeling Skills: Process modeling and simulation, Transport simulation, Computational Fluid Dynamic (CFD) Simulation, Molecular Dynamics (MD) Simulation

3. Programming Skills: C, C++, Fortran, Phython

578294 Characteristics of Phase Changing Material Nanoemulsions

Characteristics of Phase Changing Material Nanoemulsions

Artur Zych[2], Jungeun Park[1], Dinesh V. Kalaga[1,2], Robert Messinger[1], Masahiro Kawaji[1,2]

[1]Chemical Engineering Department, City College of New York, CUNY, New York, NY-10031.

[2]Mechanical Engineering Department, City College of New York, CUNY, New York, NY-10031.

*Corresponding author: Artur Zych, Ph.D., E-mail address: arturzych93@gmail.com

Phase change material (PCM) is an energy storage alternative that consists of heat storage performance due to latent heat capacity. In buildings, PCMs can be utilized to save energy in heating and cooling systems implementing it into the walls of the structure. However, an alternative is circulating the PCM throughout the building by firstly turning them into emulsions. The motive is to create nanoemulsions of these PCMs with the lowest possible viscosity to ensure pumpability while increasing the amount of PCM by mass to enhance larger latent heat storage capacity which can reduce energy cost. The concern with nanoemulsion PCMs is choosing the correct emulsifiers as well as phase changing material based on applicability, stability, and economical costs. Knowing the finer details of these substances can be a stepping stone to choosing the correct materials.

This study aims to analyze the thermal performance and molecular structures of these PCM nanoemulsions through using various surfactants, thermal cycling, and the effect of aging for a size distribution ranging from 100 to 400 nm acquired by Dynamics Light Scattering (Malvern Zetasizer ZS90). Through rheological properties and molecular structures performed by NMR spectroscopy on these PCM nanoemulsions we may see how these nanoemulsions evolve and how this can create differences in the heat transfer coefficients obtained from forced heat convection in a heated circular tube. Octadecane is used as the PCM in the dispersed phase, with the utilization of different surfactants: stearic acid and the tween & span families, within the continuous phase as water. From preliminary data acquired of octadecane at a volume fraction of 30% with Span 60 and Tween 60 emulsifiers (3 wt%) showed shear thinning effects with the presence of minute yield stresses observed at all temperatures (10 -35°C) indicating a viscoelastic region; evident by the exponentially decreasing slope of the viscoity profile with an increase in shear rate (0.01-1000 S^{-1}). The temperature dependency is undoubtedly visible as the viscosity decreases with increase temperature. Performed on 50mm parallel plate geometry with a peltier heating plate for the higher volume fraction of oil(octadecane), as well as through a cup-and-bob geometry utilizing the ARES G2 Tensiometer (TA instruments). Solution-state ^{1}H and ^{13}C NMR measurements reveal molecular-level information on the local environments, compositions, and dynamics of the oil and surfactant species. Which of whom significantly depend on the thermodynamic state, age, and size distribution of the dispersed phase. While Differential Scanning Calorimetry (TGA Q1500) shows the initial PCM emulsion melting temperature and how it changes with thermal cycling and aging. A test section consisting of 304 stainless steel tube heated uniformly and thermocouples at the inlet, outlet, and several on the tube wall to measure the temperature along the tube, the heat transfer coefficient can be acquired. This can indicate the possibly of using the PCM nanoemulsion as an applicant for thermal energy storage and transport.

Keywords: PCM, Nanoemulsion, Viscosity, NMR

563431 Evaluation of the Impact of Uncertainties of Measurement of Process Parameters in the Calculation of Incrustation Resistance in Crude Oil Heating Heat Exchangers

An important phenomenon in the operation of heat exchangers is fouling, which is characterized by the sedimentation of solid material on the surface in the walls or tubes of such equipment. The deposition can have three origins: crystallization, chemical reaction or biodeposition. It is an important phenomenon since the deposited solid material can restrict the cross-sectional area for fluid flow, causing increased pressure drop and resistance to heat transfer, limiting the performance of the equipment in energy terms and increasing energy costs and cleaning.

Fouling is a particularly complex phenomenon when it occurs in crude oil pre-heating (BPA) batteries, which have the function of supplying the oil stream, which feeds the refineries at a temperature level which is sufficient to maintain the efficiency of the atmospheric distillation processes and under vacuum. The complexity is due to the multiplicity of phenomena involved that influence and are influenced by deposition, such as activation energy of chemical reactions, solubility of asphaltenes, presence of particulates, flow velocity, among others. These parameters affect fouling, making any mathematical model applicable only to very restricted conditions, which in many cases do not represent the operational conditions of a refinery.

Nevertheless, many efforts have been made to construct mathematical models capable of being widely applied in predicting the dynamic behavior of deposition in GAPs. The great difficulty factor of the construction of these models consists of the multiplicity of phenomena that act in a synergistic way in the generation of the deposition, so that the complexity of the phenomenon is not usually fully represented by the contour conditions of the models. Thus, many of these models show deviations from the operational data obtained in industrial units. However, the analysis of the pertinence of the models is generally carried out without considering the uncertainty of the data obtained in the industrial plants, which can make difficult a more accurate assessment of the degree of approximation of reality.

This work presents a methodology to estimate the thermal resistance (R_f) measurement errors in refinery oil preheating batteries. For this, mathematical correlations were developed to estimate the error in the R_f calculation based on the temperature and flow measurement uncertainty reported by the instrument manufacturers. The methodology was implemented by calculating the error range of R_f calculations performed from available data from oil refineries. As a result, relative errors of R_f of the order of 50% were observed, indicating that some mathematical models may be considered valid if uncertainties in the measurement of operational parameters are taken into account.

556670 Hydrodynamics and Mass Transfer of Gas-Liquid Flow in a Tree-Shaped Parallel Microchannel with T-Type Bifurcations

Gas-liquid systems are widely encountered in various chemical processes. However, the process efficiency is usually insufficient in traditional equipment. As an effective approach for process intensification, micro-chemical technology has received increasing attention. It has been verified that the mass transfer efficiency of a single microchannel is much higher than that of the conventional reactors. To maintain the merit of micro-scale, the numbering-up strategy has been adopted to implement industrial throughputs. Nevertheless, the numbering-up strategy could cause negative effects on hydrodynamics and mass transfer. Therefore, the study of the dependence of hydrodynamics and mass transfer of gas-liquid two-phase flow on the operating conditions in parallel microchannels is indispensable for the optimization of design and operation of commercial microreactor.

In this study, the hydrodynamics and mass transfer of gas-liquid two-phase flow in a tree-shaped parallel microchannel constructed by T-type bifurcations are investigated experimentally. The results show that the flow distribution is corporately determined by the capillary number *Ca* at the T-type bifurcation and the hydrodynamic feedback effect of the end of branches. At a fixed absorbent concentration, there exists a critical ratio of gas flow rate to liquid flow rate for an optimal flow distribution. Before the critical point, the uniformity of flow distribution becomes better with increasing the ratio, but it would gradually deteriorate beyond the critical point. Meanwhile, the operating condition also determines the flow regime in each branch, thereby influences the mass transfer performance. Before the critical ratio, bubbly flow or slug flow occurs in the branch, and the specific surface area extends with the increase of the ratio. However, beyond the point, the flow regime evolves into the compact slug flow, the specific surface area only decreases slightly with the increase of the ratio, and the liquid slug in the branch almost disappears, consequently the mass transfer depends mainly on the liquid film surrounding the main body of bubble instead of the liquid slug. Due to the change of hydrodynamics in the microchannel, the volumetric mass transfer coefficient decreases with the increase of the gas-liquid flow rate ratio after the critical point.

In conclusion, the hydrodynamics in the tree-shaped parallel microchannel is much different between before and after critical point. Therefore, the determination of the critical point of the gas-liquid flow rate ratio could effectively avoid the negative effect and thereby maintain the excellent performance like in a single microchannel.

580020 Modeling and Experimental Study on CO_2 Adsorption in Fixed-Bed Columns: Applications to Carbon Capture and Utilization

The concerns towards the rise in carbon dioxide (CO2) concentrations in atmosphere and its impact on climate change have been increasing since past decades. While the primary source of CO2 emissions are power plants (large point sources) that burn fossil fuels to generate electricity, the cumulative emissions from vehicles and humans/animals (small but distributed point sources) can be significant. This necessitates the development of carbon capture technologies not only from flue gases but also directly from the atmosphere. Most of these technologies are based on CO2 adsorption in a fixed-bed reactor. The current work presents the modeling and experimental studies on the topic, where we developed reduced-order model to describe the transport and adsorption of CO2 in a packed-bed reactor and perform the lab-scale experiments to investigate adsorption characteristics. The reduced-order model (ROM) is expressed in multiple concentration modes capturing the small-scale physics using mass-transfer coefficients. We illustrate that ROM developed here are more accurate than traditional pseudo-homogeneous model and can be utilized for real time optimization and parametric/bifurcation studies. We also show the use of ROM in solving inverse problem to analyze the experimental data in obtaining transport/reaction parameters. Finally, we provide the physical interpretation of the various terms used in the model and discuss its utility in design/scale-up of the reactor.

575802 Thermal Diffusion in Thermogalvanic Cells Containing Polymer Electrolyte

According to annual energy flow diagrams from Lawrence Livermore National Lab, over half the energy produced in the United States is lost as waste heat. Much is low-grade heat that cannot readily be recovered by traditional thermodynamic cycles. Thermogalvanic cells are a potential solution to recovering low-grade waste heat. In contrast to thermoelectric devices that require high temperature gradients to generate power in semiconductor devices, thermogalvanic cells are like batteries that are powered by temperature gradients. Solid polymer-electrolyte-based cells could provide a flexible method to convert low-grade waste heat into electricity. Thermal diffusion, also known as the Soret Effect, is an important process in thermogalvanic cells. The temperature gradient induces a concentration gradient in the electrolyte. Theoretical prediction of the magnitude and direction of thermal diffusion is not possible despite a thermodynamic theory for small temperature gradients and a force-based theory derived for thermophoresis of colloidal particles. The missing link to understanding this phenomena may lie in a novel and interesting system that spans the gap between small molecule mixtures and colloidal dispersions, namely polymer electrolytes. We have measured thermal diffusion in a dry polymer electrolyte using two innovative approaches designed specifically for solid systems. The first approach measures the electrochemical potential of a thermogalvanic cell subjected to different thermal gradients. At steady-state, the electrochemical potential is directly related to the concentration gradient in the polymer electrolyte. We have also conducted transient current-voltage measurements of the thermogalvanic cell to determine the power that can be produced by the temperature gradient. These measurements will be corroborated with a second method based on spectroscopic measurement of concentration gradients in polymer electrolyte exposed to different temperature gradients. The results will be analyzed using a transport model based on irreversible thermodynamics in order to better understand the phenomena underlying thermal diffusion.

553034 Energy Optimization By Installation of Heat Recovery Steam Generator at Exhaust Gases of Gas Turbine Driven Compressor

This paper encompasses operational experience of **Energy Efficiency** project at ammonia plant of Engro Fertilizer Pakistan. The plant is a Becthel corporation design built up in 1967. The plant was relocated to Daharki (Pakistan) from Pascagoula (USA) in 1992. The unit capacity is 1550 MeT/Day (Ammonia Production)

There were two parallel combination of gas driven centrifugal compressor (2 x Taurus 60 Compressor Sets) installed for compression of natural gas for ammonia plant. The exhaust gases were vent into atmosphere resulted in wastage of useful energy

A study was conducted to recover this heat energy by installation of heat recover steam generator at exhaust gases of compressors. A state of the art HRSG (M/S DESCON as EPC vendor) was proposed for heat recovery. The new HRSG comprised of de-aerator, BWF pumps, flash vessel, economizer, super heater, and chemical dozing facility. The steam produced by the new HRSG was integrated with the existing steam circuit

The multiple improvisation steps have been taken to develop the operations strategy for integration of new HRSG with existing steam network. Owing to the complexity of steam and gas network, an in depth process hazard analyses and review was conducted for development of commissioning procedures and emergency handling

A flawless execution of the project was accomplished by the team. A total estimated steam produced by installation of new HRSG = 21 TPH (without auxiliary or duct burner) with pressure = 50 kg/cm2. (Based upon 100% load on both compressors) HRSG is designed for turndown ratio 3:1

The intent of this paper is to share the operation experience & to gauge the key learning's during this energy efficiency project. This energy optimization is a classic example of operating a process unit with higher efficiency and improving the overall energy index. This would be beneficial for the industries across the globe

553658 Thermoenergetic Analysis of a Steam Generation and Distribution System

A thermoenergetic evaluation of the steam generation and distribution system of the Universidad del Atlántico was carried out, using as reference for the determination of the thermal efficiency of the steam generators the "PTC Code 4.1 for steam generation units of the ASME". The evaluation of the main boiler showed a combustion efficiency of 75%, with a deficient air/fuel ratio mainly, and a thermal efficiency of 77.97 ± 1.23%. By operational adjustments in the burner, the excess of air was reduced by 91%, increasing the efficiency of combustion by 9.7% and thermal efficiency by 6%, achieving an economic saving of $ 1,312,840 COP/year and 0% of CO emissions due to incomplete combustion. The condition of the thermal resistance installed in the steam generation area, in the steam distribution and in the return of condensates was evaluated by a thermographic study.All the radiant and convective heat losses in the system were quantified, which amount to $ 253,700 COP / year.

578450 A Systematic Approach in Developing Alkaline Electrolysis Based Power to Gas Infrastructures Business Models

Renewable energies are rapidly increasing their portion in the overall energy supply. There are many advantages in the renewable energy including mitigating climate change. It is important to address the emerging issue because of the increasing penetration. One particular problem widely witnessed is the increasing mismatch between renewable energy supply and demand. A large amount of valuable energies is wasted and curtailed.

Power to gas technology is drawing an increasing attention as an only realistic solution for this because it can take advantage of redundant electricity energy into other energy sources that can be stored and used again later. Other energy storage options such as energy storage system is not desirable at the most because of high capital investment and deteriorating efficiencies. At the moment, Alkaline electrolysis is one of most established technologies that generate hydrogen from redundant renewable energy. Most attention has been given to develop specific manufacturing technologies and the corresponding high performing material. Active works on experiments in laboratory and pilot scale plant has been reported in the literature. There is much to be done for commercially competitive power to gas industry.

This work is concerned with identifying potential issues in constructing alkaline electrolysis based power to gas infrastructure in commercial scale. Particularly it focused on establishing the value chain aspects to put the actual power to gas infrastructure into action. After cutting edge global R&D status are analyzed. the impact of internal and external factors for the operation of the alkaline electrolysis power to gas value chains are evaluated. Issues are addressed for developing a decision-making framework for the operation of alkaline electrolysis supply chains under the presence of various factors. A recent South Korea case is presented to illustrate the applicability of the decision-making framework with some remarks.

578456 Design of Optimized Carbon Dioxide Distribution Networks

This paper addresses modeling and optimization in carbon dioxide distribution networks with focus on utilizing optimal usage of carbon dioxide (CO2) from natural gas processing to be further processed in urea and methanol facilities. The data for the study is obtained largely from the existing ammonia plants that used steam methane reforming (SMR) process to produce hydrogen from natural gas, CO2 removal unit to capture and release CO2 from synthesis gas, and Habor Bosch process to produce ammonia. The multiple options of technologies and different configurations for utilizing carbon dioxide leads superstructure formulation for the problem. The main objective of this work is to provide a systematic framework that enables the consideration of many alternatives for the superstructure development, and then evaluate them. Sensitivity analysis involving a wide range of economic changes including the natural gas market prices and product prices will be performed to determine effective factors on decision-making relating to maximizing profits. Furthermore, the developed framework would be beneficial for the decision maker in the petrochemical industry to examine and optimally select the most cost effective and environmentally sustainable pathway in the carbon dioxide distribution networks.

575093 Carbon Dioxide Removal from Biogas Using Amine Based Silica Sorbents

Biogas is a source of methane that can be used directly as fuel for combustion engines, gas turbines, and fuel cells as a renewable energy source after its purification. With CO_2 being the major contaminant (30–50%), its removal from methane becomes one of the critical steps in upgrading to increase the energy content of the biogas. Commercial techniques such as scrubbing and pressure swing adsorption (PSA) which uses high pressure for selective adsorption of gas molecules on solid surfaces has major disadvantages like corrosion problems, heavy use of water and/or decreased efficiencies due to the high temperatures/pressures used in the regeneration step [1]. Production of compressed natural gas (CNG) requires high pressures in order to separate CO_2 from CH_4 and compressor costs can be high [2]. Compressors contribute to more than 50% of the costs for a waste to energy processes. CO_2 adsorption using amine functionalized silica is a low pressure process and using this reduces the capital and operating expenses of compressors required for biogas upgrading. Mesoporous silica functionalized with amine groups have proven to be good adsorbents of CO_2 with high selectivity, low energy utilization and low regeneration costs [3]. APTES (3-Aminopropyl triethoxysilane) modified silica has been studied before [4] [5] but never tested for CO_2 separation with actual biogas.

In the work reported here, APTES immobilized on SBA-15 was prepared using conventional grafting techniques. Techniques including XRD, N_2 physisorption, FTIR, TPO were used for characterization. We tested a series of APTES modified SBA-15 with different loadings of amines for adsorption experiments of CO_2 at room temperature and 1 atm for a dry 50% CO_2 in He feed. Results show that with an increase in APTES loading (from 12 wt % APTES to 26 wt% APTES) the CO_2 adsorption capacity increases. The adsorption capacity of 12 wt % APTES and 26 wt% APTES was 0.07 mmol CO_2/g_{ads} and 0.85 mmol CO_2/g_{ads}, respectively. The adsorbents were regenerated at 100 ^{o}C. Preliminary CO_2/CH_4 selectivity study for a 50% He and dry CO_2/CH_4 feed in 1:1 ratio was done. It was seen that the adsorbent has a high selectivity ratio ~ 5 for the CO_2/CH_4 binary mixture. Currently, we are investigating the CO_2 separation in simulated biogas containing water. Following this the regeneration for large number of cycles will also be tested.

References:

1. Sun, Q., et al. (2015). "Selection of appropriate biogas upgrading technology-a review of biogas cleaning, upgrading and utilisation." Renewable and Sustainable Energy Reviews **51**: 521-532.
2. Jaramillo, P. and H.S. Matthews, Landfill-gas-to-Energy projects: analysis of net private and social benefits. Environ Sci Technol, 2005. 39(19): p. 7365-73.
3. Liu, X., et al. (2017). "Different CO_2 absorbents-modified SBA-15 sorbent for highly selective CO_2 capture." Chemical Physics Letters , **676**: 53-57
4. Mafra, L., et al. (2018). "Amine functionalized porous silica for CO_2/CH_4 separation by adsorption: Which amine and why." Chemical Engineering Journal **336**: 612-621.
5. Vilarrasa-Garcia, E., et al. (2015). "CO_2/CH_4 adsorption separation process using pore expanded mesoporous silicas functionalizated by APTES grafting." Adsorption-Journal of the International Adsorption Society **21**(8): 565-575.

575319 Promoted Effect of D-Mannitol on Thermal Dehydrogenation of Ammonia Borane

Increasing CO_2 emissions from burning fossil fuels put tremendous pressure on the ecosystem cycle. In this context, hydrogen is a clean energy carrier that produces water, heat, and electricity when consumed in fuel cells. Among the chemical hydrides, ammonia borane (AB, NH_3BH_3) has attracted a great deal of attention as a promising hydrogen storage candidate, owing to its high hydrogen content (19.6%), decomposition under the moderate condition, and stability at room temperature. However, relatively high operating temperature, long-induction period, low H_2 yield, and slow kinetics remain challenges for thermal dehydrogenation of AB. In the present work, d-mannitol was employed as an additive to promote AB dehydrogenation. The effect of d-mannitol was investigated over a wide range of AB weight ratio, pressure, temperature, density, and heating rate. It was found that hydrogen yield and thermal characteristics were influenced by these operating parameters. Furthemore, the hydrogen release kinetics for AB dehydrogenation is significantly enhanced by addition of d-mannitol. Along with rapid kinetics, the high hydrogen equivalent (> 2) was obtained at 1 atm and 85 ^{o}C, near PEM fuel cell operating temperatures. Finally, TPD/MS and ^{11}B MAS NMR analysis were performed to understand reaction mechanisms. The results show that our proposed method is promising for hydrogen storage and can be used in various applications equipped with PEM fuel cell.

567098 Techno-Economic Feasibility of Thermocatalytic Conversion of CO_2 into Renewable Natural Gas (RNG)

Introduction: The renewable natural gas (RNG) generation process discussed in this talk relies on the thermocatalytic Sabatier reaction:

$$CO_2 + 4H_2 = CH_4 + 2H_2O + \text{heat (165 kJ/mol)}$$

The main idea is to convert the CO_2 contained in waste streams (e.g. landfill gas, biogas) into RNG on site using electrolytic hydrogen (H_2) generated using renewable (wind, solar, hydro), low carbon footprint (nuclear), or surplus electricity.

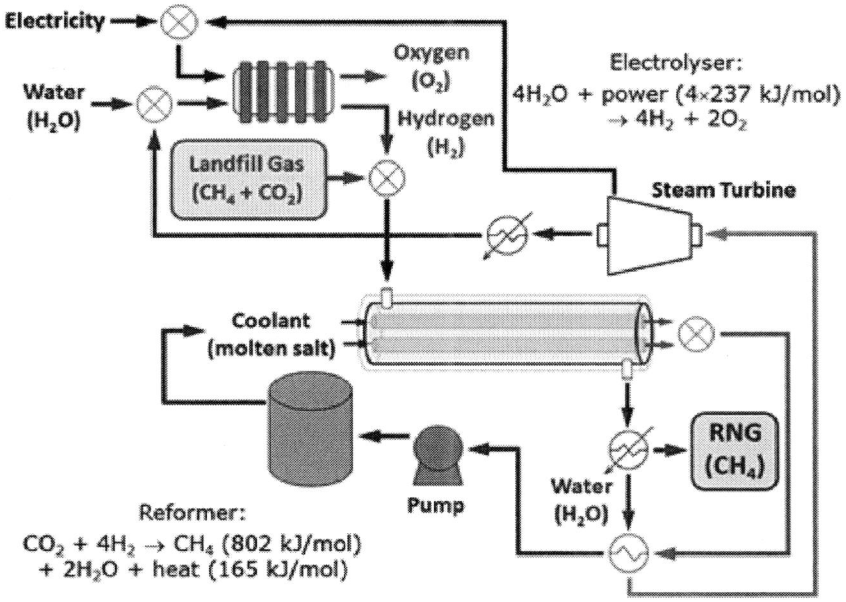

Problem Statement: Although CO_2 separation technologies are already commercially available (e.g. pressure swing adsorption (PSA) and membrane separation), the captured CO_2 has to be compressed on site, transported, and eventually stored or utilized somehow. Storing the captured CO_2 in geological formations could be prohibitively expensive. Some of that captured CO_2 can be utilized in the chemical and food industry. However, these markets are limited in their capacity to absorb megatons of separated CO_2. In addition, the capital and operating costs associated with the CO_2 compression and transportation may not be recoverable.

Innovation: The approach discussed in this talk is an innovative and sustainable alternative that goes beyond conventional CO_2 capture and storage (CCS), relying on CO_2 capture and utilization (CCU) instead. The hydrogen required for the reaction is provided by water electrolysis using renewable or low carbon footprint electricity. The water and heat generated in the process are recuperated to improve the process efficiency. Oxygen (O_2), which is a byproduct of water electrolysis can be utilized as well. With a proper postreatment, such process would produce a pipeline quality RNG, which can be directly injected into the existing natural gas infrastructure, replacing the fossil natural gas.

Results & Discussion: Technological issues related to catalysis, reactor design and system integration are briefly outlined first. The focus is on the technological and economic feasibility of this conversion pathway. A case study is included, showing the projected production cost and payout period. Design of the entire RNG production system, including upstream and downstream treatment, is discussed in detail. High-level techno economic assessments (TEA) were undertaken to evaluate the preliminary equipment, utility and financial requirements for the project. Process design utilized design software ASPEN HYSYS. Detailed TEA for the system has been conducted to analyze the project costs, profit margins and overall economics. Sensitivity analysis has been performed to test the effects of various parameters such as the electricity price. Various configurations of plant design have been considered and assessed to allow optimization.

Conclusions: The project economics is highly depended on both the RNG selling price and the price of electricity. The production cost can be as low as 10$/GJ and 15$/GJ at 4 and 8 cents per kWh electricity prices, respectively. The majority of capital and operating cost contributions are from the electrolyzer system. The maximum payout period for the system could be as low as 6 years, showing project feasibility for moderately low electricity prices.

572924 High-Pressure Hydrogen Production from Bio-Ethanol Feedstock with Fixed-Bed Chemical Looping

High-pressure hydrogen production from bio-ethanol feedstock with fixed-bed chemical looping

Decentralized hydrogen production with chemical looping hydrogen offers the use of locally available renewable energy sources and has the inherent possibility for carbon dioxide separation. The technology makes it possible to produce high purity hydrogen in one compact unit without a complex gas purification step [1]. The oxidation cycle can be performed with elevated pressure via closing the system outlet and compression of the water feed [2]. Thereby high-pressure hydrogen is generated without an additional product gas compression. Partial reduction of the oxygen carrier from Fe_2O_3 to Fe_3O_4 yields thermodynamically to a product gas only containing CO_2 and H_2O. By the subsequent condensation of steam, pure CO_2 is sequestrated. Recent experiments performed in our research group showed the ability to combine pressurized hydrogen production with the sequestration of a pure carbon dioxide stream after water condensation. The proof of concept for pure pressurized hydrogen production in combination with the sequestration of a pure carbon dioxide stream could be demonstrated. Thereby, the separation of a carbon dioxide stream with a purity of >98% could be yielded [4].

In our current research activities, a combined lab test rig is investigated in ongoing experiments based on the investigations from previous publications [3, 4]. A bioethanol feed is converted to a syngas reducing the oxygen carrier and releasing a high-pressure hydrogen stream in oxidation phase. Commercially available bioethanol is applied to investigate the effects of trace compounds on the attainable hydrogen purity in a pressurized system. Beforehand, extensive research was conducted to characterize catalysts for ethanol reforming on their capability of deriving a suitable syngas with a high share of reductive compounds in order to increase the feedstock utilization.

The in-house synthesized and pelletized iron-based oxygen carriers are characterized in a thermogravimetric analysis system as well as a fixed-bed unit for better demonstration of the impact of the exothermic and endothermic reactions on the material. Different synthesis methods – such as dry mixing, impregnation and sol-gel method are compared on the effect on the resulting material concerning cyclic stability, reactivity as well as the mechanical stability of the oxygen carrier material to satisfy the requirements of fixed-bed reactor systems. The results show a positive influence of thermal treatment at elevated temperatures (>900 °C) in combination with wet-chemical synthesis methods on the mechanic integrity for fixed-bed applications. New results of the high-pressure hydrogen production from bioethanol feed and from material development will be presented at the conference.

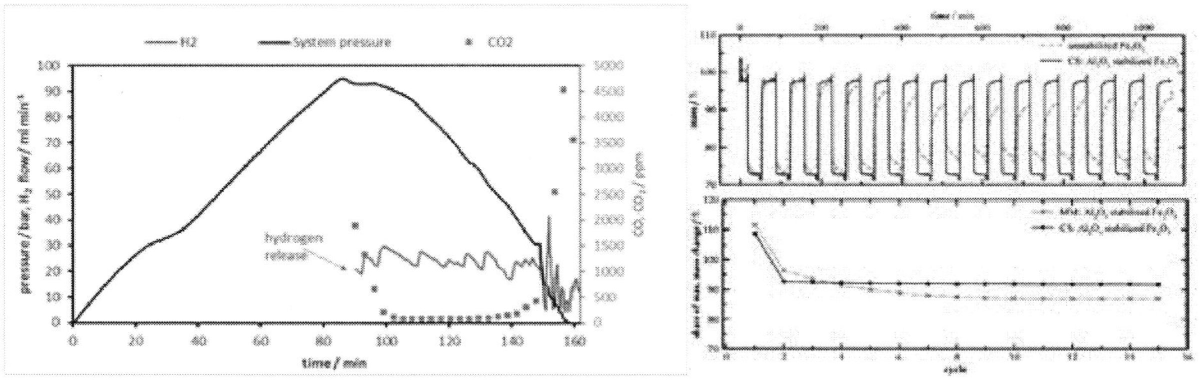

Figure 1: High pressure hydrogen release up to 95 bar and hydrogen impurities (left); Comparison of the cyclic stability of oxygen carriers gained from wet chemical synthesis, mechanical mixing and unstablized Fe as oxygen carrier (right)

[1] V. Hacker, "A novel process for stationary hydrogen production: The reformer sponge iron cycle (RESC)," J. Power Sources, vol. 118, no. 1–2, pp. 311–314, 2003.

[2] G. Voitic, S. Nestl, K. Malli, J. Wagner, B. Bitschnau, F. Mautner, and V. Hacker, "High purity pressurised hydrogen production from syngas by the steam-iron process," *RSC Adv.*, vol. 6, no. 58, pp. 53533–53541, 2016.

[3] S. Bock, R. Zacharias, and V. Hacker, "High purity hydrogen production with a 10kWth RESC prototype system," Energy Convers. Manag., vol. 172, pp. 418–427, 2018.

[4] R. Zacharias, S. Visentin, S. Bock, and V. Hacker, "High-pressure hydrogen production with inherent sequestration of a pure carbon dioxide stream via fixed bed chemical looping," Int. J. Hydrog. Energy, vol. 44, no. 16, pp. 7943-7957, 2019

554534 Biohydrogen Production from Gamma Irradiation Disintegrated Sewage Sludge: Effect of Ferrous Ions

Biohydrogen production from sewage sludge can achieve dual benefits of clean energy generation and waste management. Pretreatment process plays a vital role in biohydrogen production from sewage sludge. This study 25 kGy adopted gamma irradiation as pretreatment, explored the effect of ferrous ions (Fe^{2+}) on sludge disintegration and subsequent hydrogen production at neutral conditions. Results showed that sludge dissolution increased with the increase of Fe^{2+} dose from 0 to 1000 mg/L. Detailed analysis showed that releasement of protein prefers higher Fe^{2+} addition, while highest soluble polysaccharides was obtained at 800 mg/L Fe^{2+} addition. Indicating a proper amount of Fe^{2+} can promote the sludge disintegration by gamma irradiation. Dark fermentation was also promoted. Highest hydrogen yield of 12.23 mL/g VS_{added} was obtained from 25 kGy-200 mg/L Fe^{2+} pretreated sludge, which was 2.2 times of control test. The accumulation of volatile fatty acids was enhanced from 837.2 mg/L to 1708.0 mg/L with 1000 mg/L Fe^{2+} addition. Microbial analysis showed that the addition of Fe^{2+} significantly promoted the enrichment of *Paraclostridium* sp. and elimination of *Enterobacter* sp. Characteristics of sludge throughout the pretreatment and fermentation process were also analyzed by Excitation Emission Matrix Spectra (EEM). Interactions of sludge dissolution, hydrogen production, microbial distribution as well as Fe^{2+} concentration were also explored through Principal Component Analysis (PCA).

559514 Diffusion of 1,3,5-Triisopropylbenzene in a Serie of Al_2O_3-Y Zeolite Particles By Zero Length Column Method: From Effective Diffusion to Pore Diffusion Coefficient

The kinetics of desorption of 1,3,5-triisopropylbenzene (TiPB) in a series of Al_2O_3-Y zeolite pellets using ZLC method is reported. In this work, Five adsorbents are involved, that is, Al_2O_3, Y zeolite and three pellets synthesized by different ratios of Al_2O_3, Y zeolite and carbon black. The three pellets are named 20Y, 30Y and 40Y by the percentage of Y zeolite respectively.

In the ZLC experiment, a small sample of 20-24 mesh adsorbent pellets (equivalent spherical radius is about 0.55) are mounted in ZLC unit, except that Y zeolite is pressed to a very thin layer of crystal. Therefore, for granular materials, it's effective diffusion coefficients that are obtained in the diffusion experiments. The experiments are first carried out on Al_2O_3 materials at three different purge flow rates—100, 110, 120 sccm. The results show that the diffusion coefficients are consistent under the three purge flow rates, and thus the experimental processes are confirmed to be kinetically controlled. Therefore, ZLC experimental conditions are set as follows: Purge flow rate is 120 sccm and temperature ranges from 90 to 210 °C.

Effective diffusion coefficients calculated from ZLC desorption curve (D_{eff}) can be transformed into the effective diffusion coefficients in the pore (εD_p). The experimental result shows that under the experimental temperature conditions, the influence of the surface diffusion on εD_p can be neglected so that εD_p could be related to long straight channel diffusion coefficients (D_l) and D_l is composed of molecular diffusion (D_m) in larger channel and Knudsen diffusion (D_{kn}) in smaller channel. In this way, the correlation between D_{eff} and D_{kn} is established. The diameter of the pores in the Knudsen diffusion formula (d_{kn}) is defined to illustrate the relation between D_{eff} and D_l. As temperature increases, d_{kn} of the four samples tends to be constant. This result indicates that the diffusion processes of TiPB in the particle samples change from mesoporous control to macroporous control with increasing temperature.

560042 CFD Simulation of Flow and Heat Transfer in Fixed Beds with Axial Thermowells

Axially-centered thermowells are commonly used for monitoring temperatures in lab scale and pilot plant fixed bed tubes used in catalytic reactors for highly exothermic partial oxidation reactions. Previous analyses of the effects of including thermowells inside the tube have used effective medium models which do not represent changes in particle location and bed structure that must result from use of a thermowell.[1]

Particle-resolved simulations of fixed beds using computational fluid dynamics (CFD) is an approach that can improve our understanding of fluid flow and heat transfer by solving for flow and heat transfer around the particles, coupled to conduction, diffusion and reaction inside them. Using particle-resolved CFD, the local effects of a thermowell on the positioning of the particles and on the flow and temperature fields can be found. In this study simulation results using our validated CFD methods[2] are presented for flow with heating from the tube wall for random packings of spheres inside tubes with tube-to-particle diameter ratios (N) of 3.96, 5.96 and 7.99, respectively. Each tube-to-particle ratio (N) includes cases with inner radius (thermowell radius) to outer radius (tube radius) ratio of 0 (full bed), 0.05, 0.1, 0.2 and 0.3.

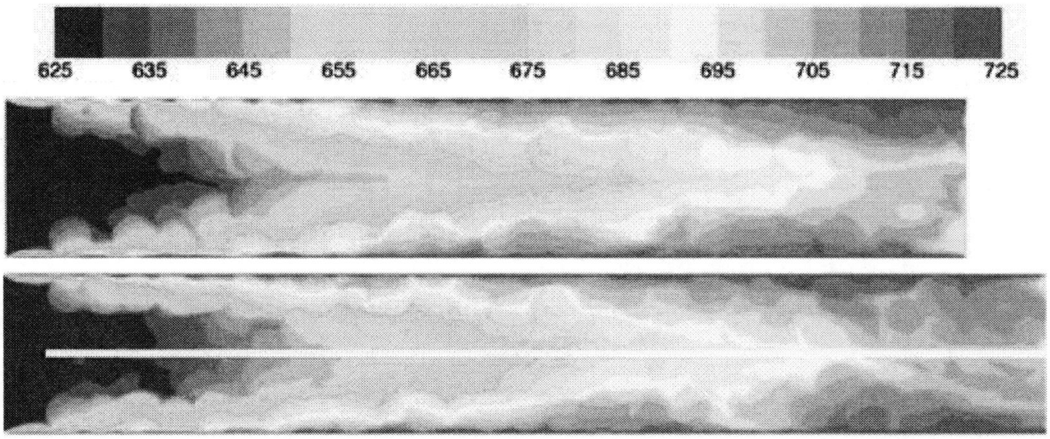

Figure 1: Temperature contours in N = 3.96 bed, comparing full bed to bed with 0.05 thermowell

Simulation results show that the presence of thermowells did change the structure, velocity field and heat transfer inside the fixed beds, mainly at the center of the tube. Pressure drop is only slightly affected. There is no clear trend with size of thermowell, the main influence on the transport is the local structure of the bed due to the arrangement of the spherical particles, which can change drastically with only a small change in thermowell size. Systematic differences are found between the centerline temperature of the tube with no thermowell, and the thermowell temperature (see Figure 1) in all cases. The changes in axial temperature occur because the configuration of particles in the bed is rearranged, even by a thin thermowell, causing changes in voidage and flow pattern and also due to axial thermowell conduction.

1. Landon V G. Temperature correction for packed bed multitubular reactors with concentric axial thermowells. Comput. Chem. Eng, 1996, 20, 475-481.

2. Dixon A G, Walls G, Stanness H, Nijemeisland M, Stitt E H. Experimental validation of high Reynolds number CFD simulations of heat transfer in a pilot-scale fixed bed tube. Chem. Eng. J. 2012, 200, 344-356.

576926 Natural Circulation of Single and Binary Gas Mixture in Very High Temperature Reactor

Natural Circulation of single and binary gas mixture in Very High Temperature Reactor
Apoorva Rudra*[a], Dinesh V. Kalaga[b], Masahiro Kawaji[b,c]
[a]Department of Chemical Engineering, City College of New York, CUNY, New York, USA.
[b]Department of Mechanical Engineering, City College of New York, CUNY, New York, USA.
[c]The Energy Institute, City College of New York, CUNY, New York, USA.

Air ingress scenario is one of the scenarios that could lead to hot spots in a very high temperature reactor (VHTR) core [1] and hence needs to be studied in detail. During air ingress, a crack/breakage in the reactor caused due to an environmental or externally initiated event leads to a pressure difference situation [2][3]. The two main gases involved are helium, which is the proposed coolant for VHTRs, and air which enters the reactor loop from outside due to the pressure difference. The air that enters into the reactor can cause oxidation of the graphite core and also induce secondary effects like increased temperatures due to an exothermic oxidation reaction. Thus, it is imperative to quantify the natural circulation behavior of these gases to get an in-depth knowledge of the air ingress scenario.

Natural circulation experiments are performed in a high-temperature and high-pressure test facility incorporating a flow channel in a graphite block simulating a prismatic core of a VHTR. Because air cannot be used owing to the presence of graphite block (possible oxidation), nitrogen is used instead in order to mimic the air flowing into the system. Mass flow experiments are conducted with the help of mass flow measurement (MFM) device, which consists of heating tape to which power is applied and two pairs of thermocouples, which are used to measure the gas temperatures and insulation temperature. The inlet and outlet gas phase bulk temperatures and surface and insulation temperatures are measured at steady state. The energy balance approach has been used to obtain the mass flow rates for different power levels.

The experimental results show interesting trends and relevant conclusions can be drawn from them. First and foremost, there is a clear dependency of mass flow rate on the working pressure because of the relation between pressure and gas density. Secondly, with respect to single gases, mass flow values of helium are found to be faster in comparison to nitrogen owing to the fact that the density of nitrogen is almost 7 times larger than helium density. A less dense gas, Helium in the present study, would be expected to move at a faster velocity. There is a drop in the mass flow if the graphite midpoint temperature (GMT) goes on increasing to a certain extent, which is not expected. Fundamentally, higher temperatures should lead to lower gas density and higher mass flow rates. However, this observation reveals the effect of two opposing properties: density and viscosity. As the helium temperature increases, even though the density decreases the dynamic viscosity increases. This leads to an increment in the flow resistance triggering the decrease in the mass flow rate. Mass flow measurements are also performed for helium-nitrogen gaseous mixture which is the actual case during air ingress scenario. There is also focus on analysis of the mean heat transfer parameters representing heat transport in the VHTR loop for different gaseous mixture concentrations.

References:

[1] Ball, S.J., et al., 2007, "Next-Generation Nuclear Plant (NGNP) Phenomena Identification and Ranking Table (PIRT) for Accident and Thermal Fluids Analysis," NUREG/CR-6944.
[2] Oh, C., Kim, E.S., Schultz, R.R., Petti, D., Liou, C.P., 2008, "Implications of Air Ingress Induced by Density-Difference Driven Stratified Flow," Proc. ICAPP '08, Anaheim, CA, June 8-12, 2008.
[3] Oh, C.H., Kim, E.S., 2011, "Air-ingress analysis: Part 1. Theoretical approach," Nuclear Engineering

and Design, 241 (1) 203–212. doi:10. 1016/j.nucengdes.2010.05.064.

572693 Structured Multiphase Reactors for Electrocatalytic CO_2 Conversion

There is a broad consensus that the chemical industry will have to make a transition to renewable feedstocks. An important route will be to use electrochemical processes, driven by electricity from sustainable sources, to produce hydrogen from water, hydrocarbons from CO_2 and water, and ammonia from nitrogen and water. This will require novel reactor designs, with special attention to the structured design of electrodes. Although there is progress in electrochemical flow reactors, scalable macro- and micro-reaction environments need to be developed involving e.g. gas-liquid electrolyte flow [1].

We examine the use of millichannels in designing multiphase reactors for electrocatalytic applications. We give a general analysis based on dimensionless numbers, followed by an illustration of this approach to catalytic packed-bed reactors. Finally, we discuss how such reactors can play an important role of increasing the number of electrocatalytic processes in the chemical industry.

Although the electrocatalysts required for CO_2 reduction receive substantial research interest, the research efforts devoted to the design of electrocatalytic reactors remains behind. Traditionally electrochemical reactors are designed as semi-2D geometries (one dimension much smaller than the other two) because of the use of plate electrodes. However, when gas bubbles are involved (e.g., due to hydrogen formation), the use of small channels might be attractive. Here we study the influence of channel size (hydraulic diameter d_h and length L), possibly filled with particles of diameter d_p, at porosity e. We consider flow at superficial velocity U of two types of fluids through these reactors, one gas-like and one liquid-like, for which we will for the moment use the physical properties of air and water at 20 ^{0}C temperature and 1 atm pressure.

The differential pressure drop necessary to reach significant convective improvement of mixing and heat and mass transfer (typically Re>100) becomes prohibitive for small channels. For example, for an empty structured channel of hydraulic diameter d_h, the pressure drop associated with wall friction can be described by the Darcy-Weisbach equation [2]. Fig. 1 shows Dp/L as a function of d_h for liquid-like (black solid line) and gas-like (black dashed line) fluids at Re = 100, assuming f = 64/Re. At this *fixed* Re, U decreases like $1/d_h$, leading to $1/d_h^3$ scaling of the required pressure drop: for d_h = 0.1 mm, the pressure drop to reach Re=100 is already substantial, approximately 10 bar/m, while for d_h = 1 mm the necessary pressure drop is a more manageable: 0.01 bar/m.

For a channel filled with catalytic particles of diameter d_p, the differential pressure drop can be described by the Ergun equation [3]. Under these conditions, for d_p = 0.1 mm, the pressure drop necessary to reach Re=100 is enormous, approximately 1000 bar/m, while for d_p = 1 mm, the necessary pressure drop is much more reasonable, approximately 1 bar/m.

Pressure drop and mass and heat transfer are not the only important consideration. We also need to consider that smaller reactors are also more prone to contaminants and agglomerates getting stuck in corners of the channel or pore space. Moreover, it is important to tune the residence time in the channel to optimize reaction yield or selectivity. The average residence time is $\tau = \varepsilon L/U$. At fixed Re, this leads to $\tau/L = \rho d_h/(\mu \mathrm{Re})$ or $\tau/L = \rho\varepsilon d_p/(\mu \mathrm{Re})$, for the two respective systems. In the regime where inertial flow enhancement becomes relevant (Re>100), reaching sufficiently *long* residence times requires sufficiently wide channels or sufficiently large particles.

Millichannels have clear advantages over traditional, non-structured reactors, in electrocatalytic processes. Analysis based on dimensionless numbers shows that the use of millichannels has important advantages over microchannels. The walls can act as electrode, while the channel configuration prevents problems with gas bubbles encountered in other types of electrocatalytic reactors.

References

[1] Walsh, F. C. & Ponce De León, C. 2018, 280, 121-148.

[2] Romeo, E., Royo, C. & Monzón, A. 2002, Chemical Engineering Journal, 86, 369-374.

[3] Ergun, S. 1952, Chem. Eng. Progress., 48, 89-94.

572576 Preparation of Expandable Graphite with a Taylor-Couette Flow Reactor

Expandable graphene (EG), which consists of stacked, flexible graphene sheets capable of expanding upon heating, has proven to be an instrumental material for a wide range of electronic applications. The production of EG has been limited to batch processes using numerous intercalating agents such as halogens, alkali metals, sulfates, nitrates, organic acids, and metal oxides. Intercalation affects the physical and chemical properties of graphene, which results in graphite material comprising hybrid properties that are a function of the intercalating material. Intercalating agents can penetrate the interlayer spacing in between graphene sheets using ultrasound or microwave irradiation, which can result in exfoliating single- and few-layered graphene sheets upon exposure to heat. High temperatures cause the expansion and gasification of the intercalating agents resulting in sufficient pressure to break the van der Waals forces that hold the graphene sheets together.

In this study, we used a batch Taylor-Couette reactor setup and an aqueous solution to expand natural graphite flakes and produce EG. Conventional Taylor-Couette setups consist of two coaxial cylinders with the inner cylinder rotating while the outer cylinder is still, creating Toroidal, or Taylor, vortices that enhance the mass transfer of the reaction medium. The Taylor-Couette setup used in this study included the rotation of the outer cylinder while the inner cylinder is still. Laminar Couette flow structure and high shear rates were achieved via the rotation of the outer cylinder, which dampened vortex formation due to the absence of centrifugal forces acting on the reaction medium. To study the effect of residence time on graphite expansion, different residence times were employed and investigated.

Our results reveal that expanding graphene sheets could be achieved in an aqueous solution using a dispersant and a stabilizing agent. In addition, the level of expansion using an aqueous solution and a Taylor-Couette reactor is comparable to commercial EG synthesized by intercalating sulfuric acid, as revealed by x-ray diffraction. More importantly, the resultant EG flakes were more structurally homogeneous than commercial EG, and residence time exceeding three hours resulted in structural defects on the EG flakes. The resulting solution could be used for electrode fabrication in lithium-ion battery setups and graphene fiber applications.

PIV Experimental Characterization of Fluid Dynamics Generated by Rotating Inner Cylinders with and without Rough wall in a Taylor–Couette Reactor

*Bin Li, Chenyang Xue, Xiaogang Yang, Guang Li**

Department of Mechanical, Materials and Manufacturing Engineering, University of Nottingham Ningbo, China

Abstract

The vortex flow in a Taylor-Couette reactor (radius ratio η=0.8, aspect ratio of 10) was experimentally studied with the help of a two-dimensional particle image velocimetry for a range of Reynolds numbers Re varying from 2498 to 4164, corresponding to turbulent vortex flow regime. The effect of the rough surface of rotating inner cylinder on the velocity field and turbulence in the Taylor-Couette reactor has been investigated. The rough wall of inner cylinder is achieved by wrapping on the smooth wall inner cylinder a perforated wire mesh with square holes, which is characterized by a mesh number of 20 and a wire diameter of 1mm. The velocity field and the turbulent field were obtained from the PIV measurement. The experimental results acquired in the TC reactors adopting a rotating inner cylinder with rough surface were compared with that in a TC reactor using a smooth rotating inner cylinder of the same diameter. It is found the Taylor–vortex was obviously deformed by using the inner cylinder with a rough surface. The highest turbulent kinetic energy was found to occur at the position of outflow boundary of Taylor vortices for both TC reactors using inner cylinder with rough surface and inner cylinder with smooth surface. However, the overall turbulent kinetic energy of flow field generated by rough surface inner cylinder is significantly increased by almost 300% in comparison with that produced by smooth inner cylinder when Reynolds number is 4164. In terms of turbulent energy dissipation rates, the results show that the maximum turbulent dissipation rate is also observed at the outflow boundaries of Tylor vortices, the overall turbulent dissipation rate of the fluid field generated by the smooth surface inner cylinder is greatly smaller than that of inner cylinder with rough surface.

Keywords: Taylor-Couette flow, particle image velocimetry, rough-wall inner cylinder, Turbulence

1.Introduction

Taylor-Couette (TC) reactor is a device typically consisting of a rotating inner cylinder and a stationary hollow cylinders in which a sequence of flow regimes can be generated by continuously changing the rotation rate of inner cylinder. Since the initial work of Taylor [1], the flow fields in TC devices have been examined by many authors using different experimental methods, such as PIV, LDV and UVP technologies. The main advantages of TC devices, compared to classical turbine stirred tank, are the changeable vortex flows varying with the rotation rate of inner cylinder and the large specific reactor surface due to small gap between the concentric inner and outer cylinders. Numerous applications of applying these advantages were reported in literatures, such as liquid–liquid extraction[2], catalytic[3], water purification[4], culture of animal cells[5], photochemical[6], micro-particle classification[7] and cultivation of microalgae[8-10].

However, most of the experimental studies of TC flow reported so far have been conducted in TC reactors using round rotating inner cylinder with smooth surface. In fact, non-smooth surfaces are widely encountered in nature and engineering applications, such as heat exchange using rough pipes. Unfortunately, as far as we know, for TC reactors, only Sodjavi, et al. [11] numerically investigated the effects of the rough wall of inner cylinder on the hydraulic resistance in turbulent Taylor-Couette flow, and they concluded such effect can be attributed to plume-ejecting phenomena and the change of near wall boundary structure, which is further dependent on the Taylor number. Although very few work has been done regarding the effect of rough wall of the rotating inner cylinder on the flow dynamics in TC reactor, the effect of configuration of inner cylinder on the Taylor–Couette flow has attracted some efforts. For example, Richter, et al. [12] measured the macro and micro mixing conditions in a continuous flow Taylor-vortex reactor using round inner cylinder equipped with horizontal ribs. They found the ribs attached on the inner rotor are able to immobilize and stabilize the vortices, and the micro mixing is obviously enhanced along with the reduced axial dispersion in comparison with the cylindrical rotors. Greidanus, et al. [13] reported the drag reduction effect of using a riblet surface on the inner cylinder. The three velocity components in the instantaneous flow field were obtained by tomographic PIV and indicate that the friction coefficients are strongly influenced by the flow regimes and vortex structures.

In this study, the rough wall of inner cylinder is achieved by adopting a square hole perforated wire mesh with a mesh number of 20 and a wire diameter of 1mm, which was firmly wrapped on the inner cylinder. In order to examine the effect of the rough wall of inner cylinder on the flow field in the Taylor-Couette reactor, the smooth inner cylinder was also studied for comparison purposes. A 2-D PIV system was used to captured the flow fields in Taylor-Couette reactors with both smooth wall and rough wall inner cylinders, including the mean velocities, fluctuation velocities, turbulent kinetic energy (TKE) and local specific energy dissipation rates, ε_{DE}.

2. Materials and methods

2.1 Taylor-Couette facility

The flow is generated between two coaxial cylinders (Fig. 1). The inner cylinder has a radius of $r_i = 40mm$ and the outer cylinder of $r_o = 50mm$. The gap between the cylinders is thus $d = r_o - r_i = 10\ mm$. The length of the inner cylinder is $L = 100\ mm$. The inner cylinder can rotate with the help of a servo driver, the rotating speed can vary from 60rpm to 100rpm. To avoid the light reflection when the laser illuminates on the inner cylinder, of which the wall is painted in black. The outer cylinder is manufactured from transparent plexiglass. Optical distortions were eliminated by enclosing the outer cylinder in a square Plexiglas box filled with deionized water, which ($\rho = 997\ kg/m^3, \nu = 1.006 \times 10^{-6} m^2/s$) was used as working fluid fully filling up the TC reactor. The Reynolds number is defined as:

$$Re = \frac{\Omega r_i d}{\nu} \tag{1}$$

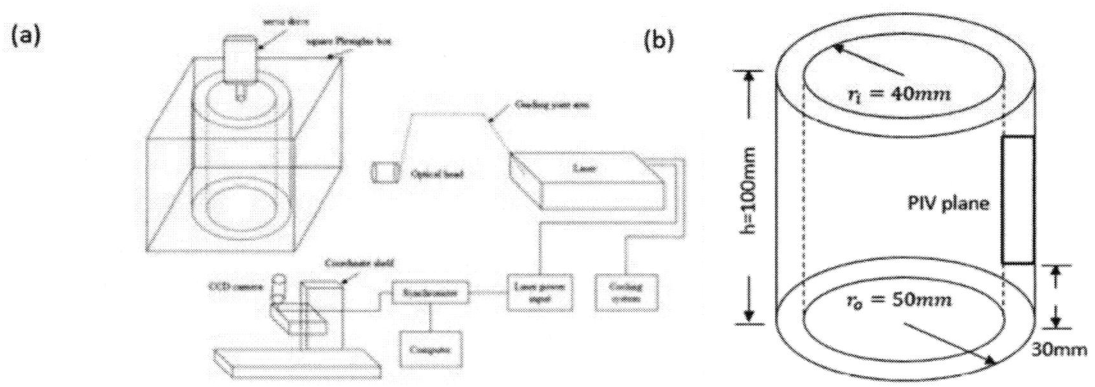

Fig. 1 a) Schematic diagram of the measuring system, b) sketch with dimensions of the Taylor-Couette reactor.

2.2 Configurations of rotating inner cylinders

This work aims to explore the effect of rough wall of inner cylinder on the dynamic behavior of Taylor-Couette vortex operating in the range of turbulent flow regime. Thus a novel inner cylinder with rough wall was designed to compare with a reference cylinder having a smooth surface. Note that for both the rough wall and smooth wall cylinders, the accompanying outer cylinder has smooth surface. Fig. 2 presents the two inner-cylinder configurations. As can be seen, there is a groove on the inner cylinder to fix the mesh. The mesh number of the mesh is 20.

2.3 PIV experiments and analysis

The 2D-PIV system provided by Dantec Dynamics (Denmark) is composed of a Nd-YAG laser system, a single frame-straddling CCD camera, a timer box and the Dynamic Studio software. The polyamide particles were used as tracer for all experimental conditions following the recommendation of Dantec Dynamics company. The images of tracer particles in the meridional plane of the gap illuminated by the laser sheet are acquired at a frequency of 5 Hz by the CCD camera, which was positioned orthogonally to the laser sheet. The time interval between the two images of each pair was carefully chosen according to the spatial resolution and rotation speed of the inner cylinder. After the PIV images were obtained, the adaptive-correlation algorithm was used to process them for yielding the velocity fields. For each experiment condition, 1000 pairs of images are recorded to ensure to reach to the steady time-averaged

Fig. 2 Picture side view of the tested inner cylinders.

velocity fields. In current work, three rotation rates were tested for both inner cylinders. (N = 60; 80;

100 rpm, Re =2498; 3331; 4164).

2.4 Evaluation of the TKE dissipation rate

According to turbulence theory, the turbulent kinetic energy dissipation rate can be calculated from Eq. (2)[14]. In Cartesian coordinate, this equation needs to be expanded in three directions for computing local dissipation rate, which consisting of 12 unknow terms formed by nine local gradients of fluctuation velocity. In general, it needs to determine all three components of instantaneous velocity from PIV measurment.

$$\varepsilon = \frac{1}{2}\nu\overline{\left(\frac{\partial u'_i}{\partial x_j} + \frac{\partial u'_j}{\partial x_i}\right)^2} \tag{2}$$

However, with 2D-PIV, only two velocity components can be measured. In order to estimate the third unknow term, the isotropy assumption from Sharp, et al. [15] will be adopted, thus the dissipation rate defined by Eq. (2) can be simplified as:

$$\varepsilon_{DE} = \nu\left\{2\overline{\left(\frac{\partial u'_1}{\partial x_1}\right)^2} + 2\overline{\left(\frac{\partial u'_2}{\partial x_2}\right)^2} + 3\overline{\left(\frac{\partial u'_1}{\partial x_2}\right)^2} + 3\overline{\left(\frac{\partial u'_2}{\partial x_1}\right)^2} + 2\overline{\frac{\partial u'_1}{\partial x_2}\frac{\partial u'_2}{\partial x_1}}\right\} \tag{3}$$

3.Results and Discussion

3.1 Flow fields

The time averaged velocity fields, measured in the meridional plane for the smooth-wall and rough-wall cases at three different rotating rates are showed in Fig. 3, in which the velocity was normalized with linear velocity of the inner cylinder $V_{tip} = 2\pi r_i N$. In order to capture two pairs of vortices, the measurement area on meridional plane has a width of 10 mm and a height of 44 mm.

From the Fig. 3a, for the smooth inner cylinder case, four and one half of vortices can be observed. The aspect ratio of each vortex approximates one. The shape of the vortex is relatively regular and resembles a circle. It can be easily seen that the region of outflow boundaries has the highest velocity value, where the fluid is ejected from the inner cylinder radially towards the outer cylinder. For the rough-wall inner cylinder, it is interesting to note that only four complete vortices are formed in the measurement area as shown in Fig.3b, and the middle pair of counter-rotating vortices are greatly stretched, thereby widening the areas of inflow and outflow boundaries. Furthermore, for the rough cases, the area of regions with normalized velocity greater than 0.16 decreases with the Reynolds number increasing. In particular, at the Reynold number of4164, the region with normalized velocity greater than 0.16 almost disappears, which maybe indicate that rough wall of inner cylinder could homogenize the velocity distribution in the annuls of TC reactor.

In order to further examine the dynamic behaviors of vortices induced by the rotating inner cylinders with or without rough surface, a sequence of six consecutive instantaneous velocity streamline fields for all experimental conditions are shown in Fig. 4. It can be found that, for smooth inner cylinder, the vortex structure is relatively stable, and the location of the inflow and outflow boundaries separating the

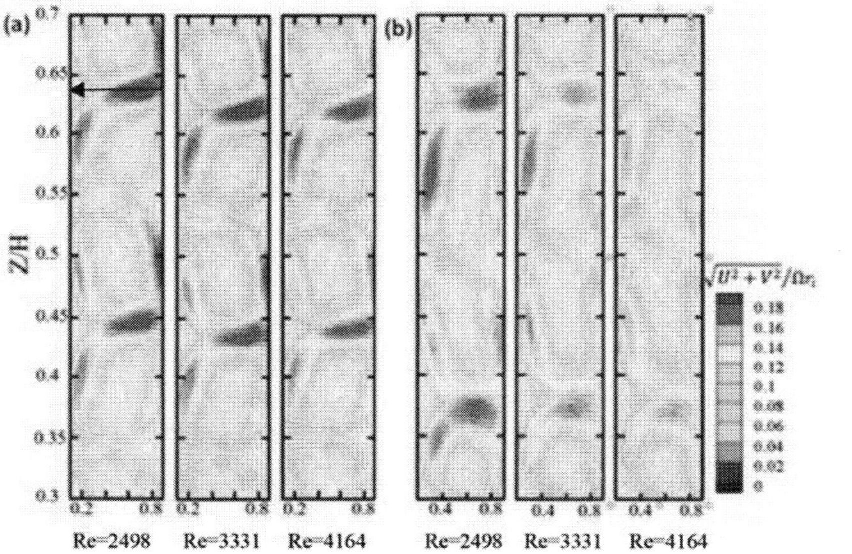

Fig. 3 Time-averaged velocity fields obtained at three Reynold numbers in TC reactor using different inner cylinder : (a) smooth-wall inner cylinder, (b) rough wall inner cylinder

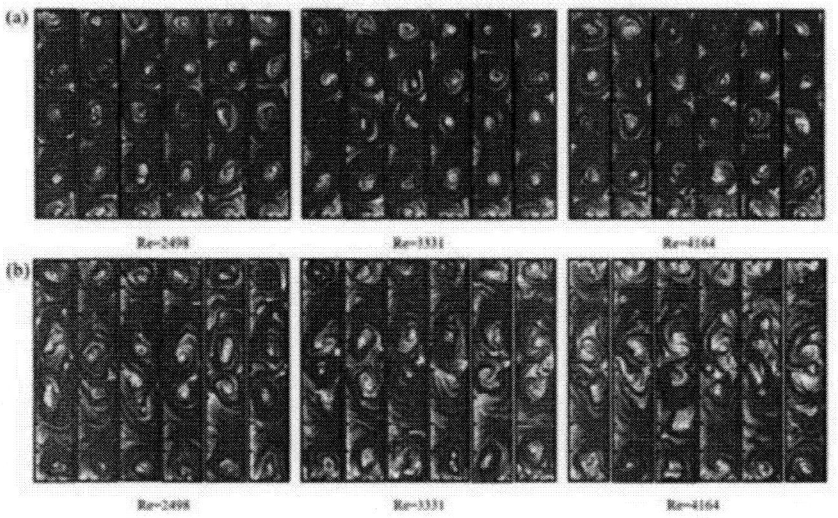

Fig. 4 Six consecutive instantaneous velocity streamline fields obtained at three Reynold numbers in TC reactor using different inner cylinders : (a) smooth-wall inner cylinder, (b) rough wall inner cylinder

adjacent vortices and the vortex centre just slightly change with time. However, in the TC reactor using rough wall inner cylinder, the vortex structure is greatly deformated, resulting in significant flow instability and variable vortex shape with time. Furthermore, both the directions and postiton of the inflow and outflow randomly varies with time. With the Reynolds number increasing, many small vortices start to appear emboding in the large scale vortices.

3.2 Turbulent kinetic energy

Fig.5 depicts the contours of time-averaged turbulent kinetic energy at various rotation speeds for smooth inner cylinder and rough inner cylinder, in which the turbulent kinetic energy is represented by k and normalized with V_{tip}^2. Combining with the figure 3, it can be seen from figure 5 that the highest k values occur at the regions of outflow boundary, and the turbulent kinetic energy sharply decreases when approaching to the center of vortex, indicating the strongly imhomgenous distribution of the turbulent kinetic energy throughout the gap. Comparing the figure 5a with figure 5b, we find that the turbulent kinetic energy produced by rough wall inner cylidner is greatly larger that of the smooth one at all three rotation speeds. Fig.6 presents the overall turbulent kinetic energy calculated by integrating the local value of k over the measurement area. At Reynolds number of 4164, the turbulent kinetic energy yielded by the rough wall inner cylinder is almost three times that of the smooth inner cylinder, which indicates that the turbulence intensity in TC flow can be massively ennhaced with help of wire mesh covering the smooth inner cylinder. Thereby it can be expected the mixing efficiency of turbulent Taylor flow could be remarkably improved by adopting inner cylinder with rough wall.

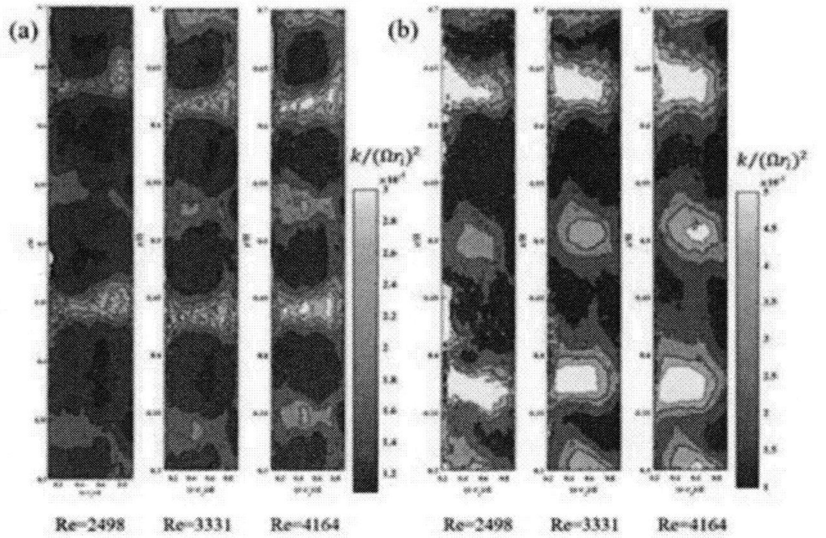

Fig. 5 Contour plots of time-averaged TKE at different Reynold numbers for: (a) smooth-wall inner cylinder, (b) rough wall inner cylinder

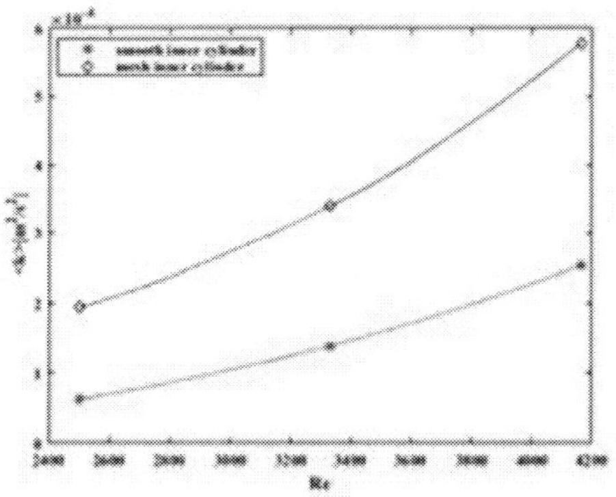

Fig. 6 Volume-averaged turbulent kinetic energy $\langle k \rangle$ as a function of Reynolds number

3.4 Local specific energy dissipation rates

Figure 7. presents the spatial distribution of turbulent kinetic energy dissipation rate obtianed for smooth inner cylinder and wire mesh inner cylinder, which resmebles the spatial distribution of turbulent kinetic energy with highest values occuring in the outflow boundaries. The difference in the local turbulent kinetic energy dissipation rate between smooth inner cylinder and rough inner cylinder is obvious. The volume-averaged energy dissipation rate indcued by the smooth inner cylinder is much smaller than rough inner cylinder as shown in Fig.8. However, with the Reynolds number increasing, the difference in dissipation rate between the two kinds of inner cylinders gradually decreases, which may indicates that the effect of rough wall on production of dissipation rate is gradually reduced as the Reynolds number increases.

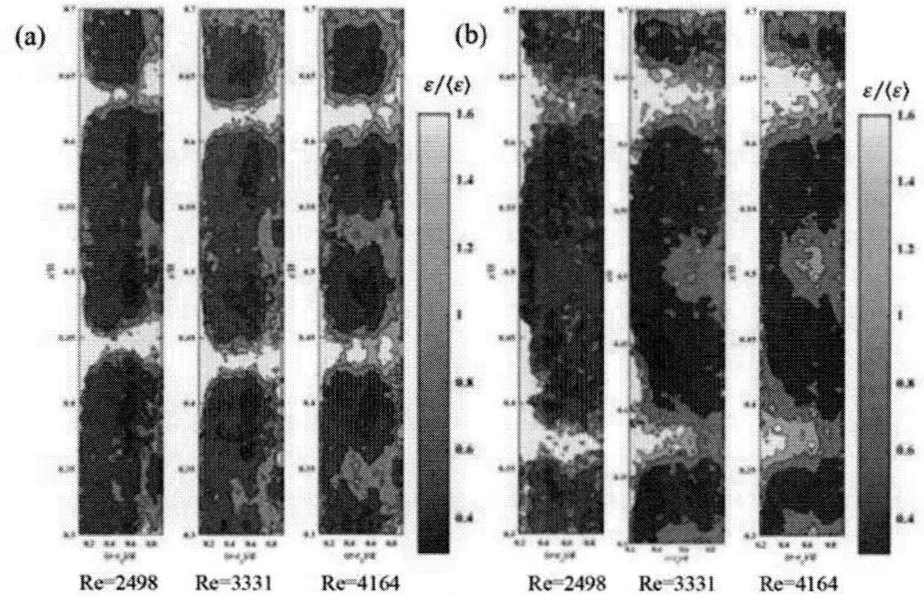

Fig. 7 The contours of local dissipation rates at different Reynold numbers for (a) smooth inner cylinder (b) mesh inner cylinder

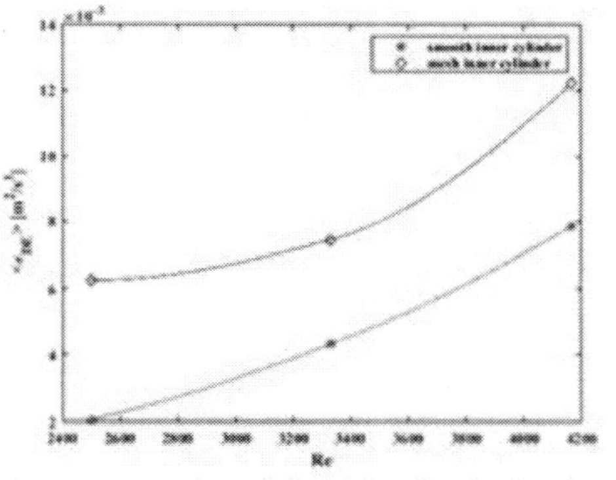

Fig. 8 Volume-averaged turbulent kinetic dissipation rate $\langle\varepsilon_{DE}\rangle$

4. Conclusions

In this work, a novel rotating inner cylinder with rough surface was proposed and used for investigating the influence of roughness of rotating wall on the flow field and vortex structure in Taylor-Couette reactor with the help of Particle image velocimetry technology. The visualization of flow structure shows that, in comparison with the smooth case, the rough-wall inner cylinder wrapped with the perforated wire mesh greatly modifies the time-averaged flow fields, significantly stretching the middle pair of vortices and contracting the bottom and top vortices. At all experimental conditions, the highest values for both turbulent kinetic energy and its dissipation occur at outflow boundaries, and the smallest values appears at the vortex centres. It's found that the turbulent kinetic energy generated by the rough-wall inner cylinder is much larger than that of the smooth inner cylinder by at least 100%.

References

1. Taylor, G. I., Statistical Theory of Turbulence. *Proceedings of the Royal Society of London. Series A, Mathematical and Physical Sciences* **1935,** *151* (873), 421-444.
2. Davis, M. W.; Weber, E. J., Liquid-Liquid Extraction between Rotating Concentric Cylinders. *Industrial & Engineering Chemistry* **1960,** *52* (11), 929-934.
3. Cohen, S.; David Moalem, M., Analysis of a rotating annular reactor in the vortex flow regime. *Chemical Engineering Science* **1991,** *46* (1), 123-134.
4. Baier, G.; Graham, M. D.; Lightfoot, E. N., Mass transport in a novel two-fluid taylor vortex extractor. *AIChE Journal* **2000,** *46* (12), 2395-2407.
5. Haut, B.; Ben Amor, H.; Coulon, L.; Jacquet, A.; Halloin, V., Hydrodynamics and mass transfer in a Couette–Taylor bioreactor for the culture of animal cells. *Chemical Engineering Science* **2003,** *58* (3), 777-784.
6. Forney, L. J.; Pierson, J. A., Optimum photolysis in Taylor–Couette flow. *AIChE Journal* **2003,** *49* (3), 727-733.
7. Ohmura, N.; Suemasu, T.; Asamura, Y., Particle classification in Taylor vortex flow with an axial flow. *Journal of Physics: Conference Series* **2005,** *14*, 64-71.
8. Miller, R. L.; Fredrickson, A. G.; Brown, A. H.; Tsuchiya, H. M., Hydromechanical Method to Increase Efficiency of Algal Photosynthesis. *Industrial & Engineering Chemistry Process Design and Development* **1964,** *3* (2), 134-143.
9. Oasmaa, A.; Elliott, D. C.; Müller, S., Quality control in fast pyrolysis bio-oil production and use. *Environmental Progress & Sustainable Energy* **2009,** *28* (3), 404-409.
10. Kong, B.; Shanks, J. V.; Vigil, R. D., Enhanced algal growth rate in a Taylor vortex reactor. *Biotechnology and Bioengineering* **2013,** *110* (8), 2140-2149.
11. Sodjavi, K.; Ravelet, F.; Bakir, F., Effects of axial rectangular groove on turbulent Taylor-Couette flow from analysis of experimental data. *Experimental Thermal and Fluid Science* **2018,** *97*, 270-278.
12. Richter, O.; Hoffmann, H.; Kraushaar-Czarnetzki, B., Effect of the rotor shape on the mixing characteristics of a continuous flow Taylor-vortex reactor. *Chemical Engineering Science* **2008,** *63* (13), 3504-3513.
13. Greidanus, A. J.; Delfos, R.; Tokgoz, S.; Westerweel, J., Turbulent Taylor–Couette flow over riblets: drag reduction and the effect of bulk fluid rotation. *Experiments in Fluids* **2015,** *56* (5), 107.
14. Hinze, J. O., Turbulence. An introduction to its mechanism and theory. McGraw-Hill, New York: 1975.
15. Sharp, K. V.; Kim, K. C.; Adrian, R., Dissipation estimation around a Rushton turbine using particle image velocimetry. In *Laser Techniques Applied to Fluid Mechanics*, Springer: 2000; pp 337-354.

577341 A Dynamic Genome-Scale Metabolic Network Model for a Novel Methanotroph-Cyanobacteria Coculture

A Dynamic Genome-scale Metabolic Network Model for a Novel Methanotroph-Cyanobacteria Coculture

Kiumars Badr, Q. Peter He and Jin Wang*

Department of Chemical Engineering, Auburn University, Auburn, AL, 36849, USA. *Email: wang@auburn.edu

Biogas is comprised primarily of methane (50%~70%) and carbon dioxide (30% ~50%), which can be produced from various waste sources, including landfill material, animal manure; wastewater; and industrial, institutional, and commercial organic wastes. EPA estimates that currently US biogas production potential is 654 billion cubic feet per year, which could displace 7.5 billiongallon of gasoline [1]. It is clear that biogas has immense potential as a renewable feedstock for producing high-density fuels and commodity chemicals. However, the utilization of biogas represents a significant challenge due to its low pressure and presence of contaminants such as H_2S, ammonia, and volatile organic carbon compounds. To tap into this immense potential, effective biotechnologies that can co-utilize both CO_2 and CH_4 are needed.

In our previous work, we have demonstrated that metabolic coupling of methane oxidation to oxygenic photosynthesis can be a highly efficient way to recover the energy and capture carbon from biogas [1, 2]. However, development of multi-organism platforms for commercial biogas conversion present significant challenges which center around our ability to control function and composition of species in the Coculture. An essential tool for the optimization, design and analysis of the coculture based biogas conversion is the development and validation of kinetics models that can accurately describe and predict the co-culture growth under different conditions [3]. To this end, using *Methylomicrobium buryatense - Arthrosipira platensis* as the model coculture system, we have developed an unstructured model to capture the growth dynamics. Specifically, Monod-like models were developed to capture coculture growth. Two sources of substrate were considered in the model: gas transferred from gas phase and gas produced *in situ*. In addition, we rely on the fitted maximum cell growth rate for both strains to capture other potential interactions. Using designed experiments and the developed model, we clearly demonstrated that the synergistic effect within the coculture cannot be fully explained by the *in situ* substrate exchange, and there must be other "metabolic links" to explain the significantly enhanced cell growth of both strains in the coculture.

In order to test different hypothesized "metabolic links" between the Methanotroph and cyanobacteria in the Coculture, in the work, we developed a structured, dynamic genome-scale metabolic network model for the Coculture system. These GEMs, especially the dynamic GEMs, offer a comprehensive picture of cellular metabolism and serve as a bridge that can better link the work of microbiologists and engineers in understanding and optimizing complex cellular metabolism. By integrating the available knowledge on each strain with data obtained in our own experiments, we use DFBAlab to implement the dynamic GEM for the Coculture [4]. For the dynamic GEM, besides the GEMs for each individual strain, the key inputs to the DFBA model are the uptake kinetics for different substrates. In this work, the substrate update rates are provided from the unstructured dynamic model we already developed. In addition, the product secretion rates and cell growth rates predicted by the validated unstructured model were used to validate the coculture dynamic GEM. Although the experimental measurements to validate the model are mainly limited to cell growth, substrate consumption and product secretion rates for each organism in the coculture, the overall dynamic trajectory (both measured low-frequency samples and unstructured model predicted high-frequency samples) offers significantly more power in validating the model, compared to a few steady state conditions. The validated coculture GEM enables the testing of different hypotheses on potential "metabolic links" within the coculture.

Reference:

[1] Badr, K., Hilliard, M., He, Q.P., & Wang, J., "Understanding the Stability and Robustness of a Methanotroph-Cyanobacterium Coculture through Kinetic Modeling and Experimental Validation." Annual AIChE Meeting. Pittsburgh, PA (2018).

[2] Badr K, Roberts N, Hilliard M, He QP, Wang J. "Photoautotroph-Methanotroph Coculture – A Flexible Platform for Efficient Biological CO_2 – CH_4 Co-utilization," Dynamics and Control of Process Systems, including Biosystems - 12th DYCOPS. (2019); accepted.

[3] Almquist J., M. Cvijovic, V. Hatzimanikatis, J. Nielsen, M. Jirstrand, "Kinetic models in industrial biotechnology – improving cell factory performance", Metab. Eng., 24 (2014), pp. 38-60.

[4] Gomez JA, Höffner K, Barton PI. DFBAlab: a fast and reliable MATLAB code for dynamic flux balance analysis. BMC Bioinformatics. (2014),15:409.

567780 Kinetics and Reactor Design for Pyrolytic Treatment of Petroleum-Contaminated Soils

Scope and Significance

While marine oil spills from offshore platforms or tankers attract most of the public attention, the vast majority of oil spills (about 98%) occur on land. Every year about 10-25 million gallons of petroleum products are spilled mostly from pipelines and fixed facilities. Petroleum hydrocarbons pose long term threats to groundwater quality, inhibit the germination and root elongation rates of plants and are extremely toxic to important soil microbes. Moreover, polycyclic aromatic hydrocarbons (PAH) present in petroleum can damage the immune systems of aquatic organisms and wildlife.

Current remediation methods for petroleum-contaminated soils are either relatively slow or have unintended consequences in the form of soil damage and high-energy usage. Even worse, some processes such as aerobic bioremediation can activate toxic hydrocarbons and transform them to more noxious byproducts such as PAH derivatives. For all these reasons, there is still a pressing need for more efficient and *sustainable remediation* of petroleum-contaminated soils.

Pyrolysis is receiving increasing attention as an on-site remediation approach because of its potential to rapidly and reliably remove total petroleum hydrocarbons (TPH) with lower energy requirements and better post-treatment soil fertility than other *ex situ* thermal remediation approaches [1-3]. Pyrolytic treatment of contaminated soils at 420°C with only 15-min residence time in a continuous reactor reduced TPH by 99.9%, lowered the total PAH concentration by almost 95%, and restored soil fertility to 98% of the clean soil level [3]. These results clearly suggest that pyrolysis has the potential for improved ecosystem restoration following remediation compared to traditional thermal technologies. However, the same study also revealed potential tradeoffs between pyrolytic treatment intensity, soil detoxification efficacy and soil fertility restoration. Specifically, treatment at 470°C for 15- or 30-min reduced soil fertility to 51% and 39% of the clean soil level, which was only marginally higher that the fertility of the contaminated soil.

The existence of an optimal treatment intensity for contaminated soils underscores the need for mechanistic insight into the pyrolysis process and for a deeper understanding of how the detoxification of contaminated soil and the restoration of soil fertility are affected by (a) the properties of both the soil and the petroleum crude, and (b) the operating conditions of pyrolysis reactors.

Kinetics and Reactor Design for Pyrolytic Treatment of Petroleum-Contaminated Soils

Taking the first steps towards a rational design of pyrolysis reactors, we have combined mathematical techniques and advanced analytical measurements to:

1. Develop a full kinetic model for the pyrolytic treatment of soils contaminated with petroleum hydrocarbons. The new model describes all thermally-induced transformations of soil components (e.g. clays, carbonates, or SOM), the evaporation of light hydrocarbons and the pyrolysis of the heavier hydrocarbon fractions.
2. Develop a mathematical model describing the operation of continuous reactors (rotary kilns) that will be used for *ex situ* treatment of petroleum-contaminated soils. This dynamic model will then be used to estimate the optimal (a) pyrolysis temperature, (b) reactor residence time, and (c) gas flow rates that will achieve a specified TPH reduction at the minimum energy cost for any given throughput of contaminated soil.

Two clean soils with different compositions were spiked with oil crudes to prepare petroleum-contaminated soils. Thermogravimetry with evolved gas analysis (TG-IR and TG-MS) was used first to identify the mineral transformations taking place as the clean soils were heated in an anoxic atmosphere. The same analytical techniques were applied to study the two distinct processes occurring during the pyrolytic remediation of contaminated soils. Desorption of the lighter hydrocarbons is the dominant process for temperatures between 150 and 350°C. Thermal cracking reactions dominate in the 350-500°C range releasing pyrolysis products (like hydrogen, methane, or olefins) and leading to the formation of a carbonaceous solid material (char) that remains in the treated soil [2]. It is important to note that hydrocarbon desorption and pyrolysis overlap with the temperature ranges over which soil mineral transformations take place.

To develop our kinetic model, we assumed that the soil consists not only of inert components but also of "reacting pseudo-components" (e.g. clays, carbonates, SOM) that release either water or carbon dioxide as the soil is heated. Weight losses due to water or carbon dioxide release accounts for more than 95% of the weight loss observed during heating of clean soils. Contaminated soils contained additional pseudo-components that account for the weight losses observed due to the desorption of light hydrocarbons and the pyrolysis of the heavy petroleum fractions to yield gaseous products and char (coke). Using the distributed activation energy approach (DAE) approach, each pseudo-component was assumed to consist of a large number of species that react with activation energies distributed according to Gaussian probability density functions (PDF). To obtain the kinetic parameters, an optimization problem was formulated by fitting model predictions to experimental thermogravimetric data. Solving this nonlinear constrained optimization problem with an interior-point algorithm, the expected activation energies and variances of all pseudo-components were estimated, together with the preexponential factors for all the reactions and the initial mass fractions of all pseudo-components.

Excellent fits to the experimental data were obtained using four reacting pseudo-components for the clean soils. Three pseudo-components were required to model clay dehydration and dehydroxylation processes that occur over a wide range of temperatures and generate different multi-peak water release patterns for each soil. One pseudo-component was enough to model carbon dioxide release from either carbonate decomposition reactions or soil organic matter (SOM) pyrolysis. Two additional pseudo-components were required to model hydrocarbon desorption and pyrolysis.

The kinetic models were then used to simulate the operation of non-isothermal rotary kiln reactors that will be used for *ex situ* treatment of petroleum-contaminated soils. To preserve anoxic conditions, the reactors were heated externally while a stream of nitrogen gas flowed through the rotary kiln to sweep the desorbing hydrocarbons and pyrolysis products. A finite stage or mixing cell model with separate solid and gas phases was used to describe the operation of the pyrolysis reactor.

Model predictions were compared to experimental data we obtained with a continuous rotary kiln reactor [3]. The TPH reductions predicted by the model agreed very well with the experimental data for all the conditions considered in our earlier study: pyrolysis temperatures, 370, 420 and 470°C; residence times 15, 30 and 60 min; solid flow rates 7, 14, and 25 lb/h.

More importantly, however, our model was able to explain the trade-offs observed in our earlier study between pyrolysis treatment intensity, soil detoxification and soil fertility restoration [3]. The clay components of the soil used in [3] undergo significant dehydration between 350 and 450°C, a range that overlaps with that of pyrolysis reactions. As contaminated soil moves through the reactor zone kept at 420°C, the reaction rate for the heavy hydrocarbon fraction is fast enough to achieve almost complete conversion within 15 or 30 min. But, the same conditions are not severe enough to completely remove the water from the clay components of the soil. If the soil is treated at 470°C, on the other hand, the clay components are almost completely dehydrated, causing irreversible soil damage and loss of fertility.

By emphasizing the importance of soil composition and by raising the possibility of trade-offs between detoxification and fertility of treated soils, this study has significantly advanced our fundamental understanding of the *multiple* and *concurrent*physicochemical processes taking place during the pyrolytic treatment of petroleum-contaminated soils. Theoretical kinetic and reactor models are essential for designing pyrolysis reactors optimized for specific petroleum-soil systems, enabling robust operation and facilitating acceptance of this novel technology by regulatory agencies and stakeholders.

References

1. Vidonish, J. E.; Zygourakis, K.; Masiello, C. A.; Gao, X.; Mathieu, J.; Alvarez, P. J., *Environmental Science & Technology* **2016,** *50,* (5), 2498-2506.
2. Vidonish, J. E.; Alvarez, P. J. J.; Zygourakis, K., *Industrial & Engineering Chemistry Research* **2018,** *57,* (10), 3489-3500.
3. Song, W.; Vidonish, J. E.; Kamath, R.; Yu, P. F.; Chu, C.; Moorthy, B.; Gao, B. Y.; Zygourakis, K.; Alvarez, P. J. J., *Environmental Science & Technology* **2019,** *53,* (4), 2045-2053.

573966 Well Placement Optimisation in a Heterogeneous Petroleum Reservoir with Geological Uncertainty

ABSTRACT

Huge investments are required in the development of oil and gas fields; this implies that decisions at any stage of the development process must be aided by sound and proven mathematical-oriented methods, because the use of intuitive engineering judgement alone cannot guarantee sustainable profitability over long periods [1]. This presents an ever-increasing motivation to apply optimisation techniques for the enhancement of petroleum production from both planning and operational/scheduling perspectives. The planning aspect typically involves activities such determining the number and locations of wells, drilling the wells and infrastructure installation. On the other hand, the operational/scheduling aspect comprises of well rate allocation, flow scheduling and routing problems [2]. Incorporating optimisation at the planning stage is a pre-emptive step that mitigates subsequent operational difficulties over the field's lifetime. However, embedding optimisation in a planification problem can be more challenging compared to production and injection scheduling optimisation in an oil and gas field. Although many studies [3, 4] have addressed the operational/scheduling aspect of the field development problem, this work is premised on the proposition that, jointly optimising the planning and scheduling problem better enhances field operability.

The uncertainty of subsurface geology is another constraint to take into consideration when optimising field productivity [5]. To capture the uncertainties in the subsurface geological (reservoir) model, geostatistical realisations of the most accurate reservoir representation are obtained using available information (permeabilities, porosities, fluid saturation) [6]. These uncertainties in the model are often reflected in the diverse outcomes of well productivities and ultimately the field's Net Present Value (the objective function) [7, 8]. Given the numerous number of possible realisations, it is impractical to consider all realisations in the optimisation; hence, it is important to apply classical reduction methods to ensure computational feasibility.

In this work, we specifically address the well placement planning and rate control optimisation problem under geological uncertainty. The difficulty of this problem is characterised by the presence of discrete variables, nonlinear and nonconvex objective function and constraints. Increased complexity arises from the computationally demanding functional evaluations from full-scale reservoir simulation, which is done in ECLIPSE®. We also exploit the MATLAB® Reservoir Simulation Toolbox – MRST® [9] for interfacing with ECLIPSE® and performing derivative-free optimisation tasks. The developed workflow is applied to a synthetic case study, for which it is demonstrated using several operational scenarios that using optimally selected realisations reduces computational time while providing best operating well locations. A comparative analysis of the performance of different derivative-free optimisation algorithms is also presented.

REFERENCES

1. Epelle, E.I. and Gerogiorgis, D.I., Optimal Rate Allocation for Production and Injection Wells in an Oil and Gas Field for Enhanced Profitability. *AIChE J.,* **in press** (DOI: 10.1002/aic.16592).
2. Tavallali, M.S. et al., 2014. Well placement, infrastructure design, facility allocation, and production planning in multireservoir oil fields with surface facility networks. *Ind. Eng. Chem. Res.*, 53(27), 11033-11049.
3. Gunnerud, V., Conn, A. and Foss, B., 2013. Embedding structural information in simulation-based optimization. Comput. *Chem. Eng.,* 53, 35-43.
4. Kosmidis, V.D., Perkins, J.D. and Pistikopoulos, E.N., 2004. Optimization of well oil rate allocations in petroleum fields. *Ind. Eng. Chem. Res.,* 43(14), 3513-3527
5. Li, L., Jafarpour, B. and Mohammad-Khaninezhad, M.R., 2013. A simultaneous perturbation stochastic approximation algorithm for coupled well placement and control optimization under geologic uncertainty. *Comput. Geosci.*, 17(1), 167-188.

6. Wang, H., Echeverría-Ciaurri, D., Durlofsky, L. and Cominelli, A., 2012. Optimal well placement under uncertainty using a retrospective optimization framework. *SPE J.*, 17(01), 112-121.
7. Bellout, M.C. et al., 2012. Joint optimization of oil well placement and controls. *Comput. Geosci.*, 16(4), 1061-1079.
8. Gerogiorgis, D.I., Georgiadis, M., Bowen, G., Pantelides, C.C. and Pistikopoulos, E.N., 2006. Dynamic oil and gas production optimization via explicit reservoir simulation. *Comput. Aided Chem. Eng.*21, 179-184.
9. Lie, K.A., 2014. An introduction to reservoir simulation using MATLAB: user guide for the Matlab Reservoir Simulation Toolbox (MRST). *Sintef ICT*, Trondheim, Norway.

569460 Modeling of Thermal Cracking of Natural Gas Liquids: Optimization with Support Vector Machine-Based Constraint Handling Scheme for Stiff ODEs

Rapid increase in shale gas production in the Appalachian Basin (Marcellus and Utica Shales) for the past 5 years has led to a significant growth in natural gas liquids (NGLs) production as well as a projected increase of these petrochemical feedstocks in the upcoming years [1, 2]. In this perspective, many existing ethylene crackers have expanded capacity and new crackers are becoming online to benefit from this unique opportunity [3]. Hence, the mathematical modeling and optimization of this process emerges as a necessity so as to determine the optimal operating conditions and the reactor configurations for the steam cracker, in such a way that the profit from thermal cracking of NGLs is maximized.

In this work, we are demonstrating (i) the comprehensive first-principles modeling of an adiabatic propane steam cracking process as a 1D plug flow reactor with coking effects [4] and (ii) the data-driven optimization of this reactor model using the ARGONAUT [5-6] framework. The mathematical model for thermal cracking includes the reaction rates expressed with Arrhenius equation, as well as the ODEs for continuity, energy, and momentum balances. The kinetics of thermal cracking of propane is adapted from Sundaram and Froment (1979) and the cracking model is validated using industrial data [7]. This simulation model is later used in the data-driven optimization phase to collect input-output data. However, the stiffness in the reactor models poses an additional complexity in retrieving the optimal operating conditions for the cracking process, especially when the data-driven optimization strategies strongly rely on collecting numerically stable samples from the reactor simulator. To this end, a novel Support Vector Machine (SVM)-based filtering approach is applied to eliminate candidate sampling points that are susceptible to the stiffness in the model and that will result with an abrupt termination of the cracking simulation due to high pressure drop. The SVM classifier identifies and removes the numerically unstable candidate points *a priori* to simulation execution by formulating an optimal separating hyperplane between stable/unstable points. This enables the data-driven algorithm to consistently converge to a profitable and feasible reactor configuration for the propane cracking process with an improved objective function value.

References

[1] Energy Information Administration. Today in Energy: Appalachia, Permian, Haynesville drive U.S. natural gas production growth, August 2018. URL https://www.eia.gov/todayinenergy/detail.php?id=36934. (accessed 04.10.2019).

[2] Energy Information Administration. Short-term energy outlook, April 2019. URL https://www.eia.gov/outlooks/steo/. (accessed 04.10.2019).

[3] Energy Information Administration. U.S. ethane consumption, exports to increase as new petrochemical plants come online, February 2018. URL https://www.eia.gov/todayinenergy/detail.php?id=35012. (accessed 02.26.2018).

[4] O. Onel. (2017). Advances in Modeling, Synthesis, And Global Optimization of Hybrid Energy Systems Toward the Production of Liquid Fuels and Olefins. Princeton University.

[5] F. Boukouvala and C. A. Floudas. (2017). ARGONAUT: AlgoRithms for Global Optimization of coNstrAined grey-box compUTational problems. *Optimization Letters*, 11(5), 895-913.

[6] B. Beykal, F. Boukouvala, C. A. Floudas, N. Sorek, H. Zalavadia, E. Gildin. (2018). Global optimization of grey-box computational systems using surrogate functions and application to highly constrained oil-field operations. *Computers & Chemical Engineering*, 114, 99-110.

[7] K. M. Sundaram and G. F. Froment. (1979). Kinetics of coke deposition in the thermal cracking of propane. *Chemical Engineering Science*, 34, 635-644.

574887 Towards a Novel Energy Price Predictive Framework: The Texas A & M Energy Price Index

The energy markets are sensitive and volatile to technological breakthroughs, changes in monetary policy, or major global challenges. Moreover, various governmental agencies, political and commercial organizations, think tanks as well as researchers and academics around the globe, consider various energy policies and their effects in an attempt to deal with the increasing concerns in energy independence, energy scarcity, energy sustainability, and pollution caused by the utilization of energy. Also, in the current environment in which all strategic political and commercial decisions and policies are assessed in economic terms, it is of outmost importance to determine the cost of energy precisely so as to evaluate their effectiveness. Considering the fact that energy affects every single individual and entity in the world, it is crucial to quantify "the unit price of energy" accurately, and study how it evolves with respect to time, major breakthroughs, challenges, energy and monetary policies. Given the lack of such a tool, we develop a novel framework, the Texas A&M Energy Price Index (EPI) that can be used as a benchmark to calculate the average price of energy to the end-use consumers in the United States.

The complicated energy landscape of the United States is carefully depicted to determine the products that are directed to the end-use sectors of the U.S. economy [1-5]. The total energy demand of these products, together with their prices, serve as the backbone of the Energy Price Index [6-10]. Since several key components (e.g. prices, demands) of the Energy Price Index have a lag between two and three months, we introduce a rolling horizon based parameter estimation model to approximate the current value of the Energy Price Index. The predictive methodology is tested on 120 months to show the accuracy of the developed framework in the prediction of the current value of the Energy Price Index. Energy price sub-indices for the transportation, residential, commercial, and industrial sectors are also developed. Some of the potential applications of the proposed index in the areas of economics, finance, engineering, law, and policy are demonstrated.

References:

- Elia, J. A.; Baliban, R. C.; Xiao, X.; Floudas, C. A. Optimal energy supply network determination and life cycle analysis for hybrid coal, biomass and natural gas to liquid (CBGTL) plants using carbon-based hydrogen production. Computers and Chemical Engineering 2011, 35, 1399-1430.
- Elia, J. A.; Baliban, R. C.; Floudas, C. A. Nationwide energy supply chain analysis for hybrid feedstock processes with significant CO2 emissions reduction. AIChE Journal 2012, 58 (7), 2142-2154.
- Elia, J. A.; Floudas, C. A. Energy Supply Chain Optimization of Hybrid Feedstock Processes: A Review. Annual Review of Chemical and Biomolecular Engineering 2014, 5 (1), 147-179.
- Onel, O.; Niziolek, A. M.; Floudas, C. A. Integrated biomass and fossil fuel systems towards the production of fuels and chemicals: state of the art approaches and future challenges. Current Opinion in Chemical Engineering 2015, 9, 66-74.
- Floudas, C.A.; Niziolek, A. M.; Onel, O.; Matthews, L.R. Multi-Scale Systems Engineering for Energy and the Environment: Challenges and Opportunities. AIChE Journal 2016, 62 (3), 602-623.
- Energy Information Administration (EIA), Annual Energy Outlook, 2019
- Energy Information Administration (EIA), Monthly Energy Review, 2019
- Lazard, Levelized Cost of Energy Analysis Version 12.0, 2018
- Bureau of Labor Statistics (BLS), Producers Price Index, 2018

U.S. Department of Energy (DOE), Alternative Fuels Data Center, 2018

575666 Intensification of Biorefinery Operations Using Building Blocks

Bioenergy is an attractive form of energy since it requires renewable sources that can be produced locally and used as raw materials for its production. One of the possible routes for bioenergy production is through lignocellulosic biorefineries, which utilize various biomass sources, e.g. corn stover, wood, rice straw and forest wastes, to produce biodiesel, steam and power in addition to pulp and paper [1]. Increasing awareness towards sustainable and modular technologies highlights the need for novel processing pathways leveraging on process intensification principles that can yield drastic improvements in several metrics, including economics, energy consumption and chemical footprint [2]. Recent developments in process intensification resulted in development of Integrated biorefinery concept with two new processing pathways: (i) pre-extraction of hemicellulose from biomass and its conversion to ethanol, and (ii) production of syngas from gasification of black liquor [3]. However, there exist further unexplored opportunities for intensification of the biorefineries to enhance the reaction yield and mass/heat transfer within the several parts of the process, e.g. digestion of pulp, evaporation of black liquor, etc. [4].

Recently, a new method for systematic process intensification, which relies on building block-based representation of chemical process flowsheets, has been introduced [5]. These building blocks can be integrated to form a two-dimensional grid embedding numerous non-trivial process alternatives [6-9]. Each block can be assigned with several physicochemical phenomena, e.g. reaction, vapor-Liquid equilibrium, solid-liquid equilibrium, etc., combination of which can yield various intensification pathways. In this work, we extend this novel representation methodology to various operations that take place in a biorefinery. Pulping which occurs in a pulp digester lies at the heart of a pulp mill and, here, we develop a new representation based on fundamental building blocks to represent the pulp digestion process. Mass transfer boundary layer is represented by the boundary shared by two neighboring blocks in different phases and digestion process is represented within the interior of the block. This building-block based representation is modeled using a mixed-integer nonlinear programming (MINLP) problem so as to minimize the energy consumption and increasing the yield of the process while searching for several intensified pathways. The proposed approach is demonstrated through several case studies focusing on the conventional Kraft pulping process, Kraft pulping process with pre-extraction of hemicellulose, Kraft pulping process with pre-extraction of hemicellulose and short fibers.

[1] Huang, H., Lin, W., Ramaswamy, S., Tschirner, U. (2009) Process Modeling of Comprehensive Integrated Forest Biorefinery – An Integrated Approach. *Applied Biochemistry and Biotechnology, 154*, pp. 205-216.

[2] Tian, Yuhe., Demirel, S.E.., Hasan, M.M.F., Pistikopoulos, E.N., (2018) An overview of process systems engineering approaches for process intensification: State of the art. *Chemical Engineering and Processing - Process Intensification, 133*, pp. 160-210.

[3] Aden, A., Ruth, M., Ibsen, K., Jechura, J., Neeves, K., Sheehan, J., Wallace, B., Montague, L., Slayton, A., Lukas, J., (2002) NREL Report TP-510-32438. Lignocellulosic Biomass to Ethanol Process Design and Economics Utilizing Co-Current Dilute Acid Prehydrolysis and Enzymatic Hydrolysis for Corn Stover.

[4] Huang, H., Ramaswamy, S., Al-Dajani, W.W., Tschirner, U. (2010) Process modeling and analysis of pulp-mill based integrated biorefinery with hemicelluloses pre-extraction for ethanol production: A comparative study. *Bioresource Technology 101*, pp. 624-631.

[5] Demirel, S.E., Li, J., Hasan, M.M.F., (2017) Systematic process intensification using building blocks. *Computers and Chemical Engineering, 105*, pp. 2-38.

[6] Li, J., Demirel, S. E., Hasan, M. M. F., (2018) Process Synthesis Using Block Superstructure with Automated Flowsheet Generation and Optimization. *AIChE Journal, 64(8)*, pp. 3082-3100.

[7] Li J., Demirel S.E., Hasan M.M.F. (2018) Process Integration using Block Superstructure. *Industrial &*

Engineering Chemistry Research, *57(12)*, pp. 4377–4398.

[8] Li, J., Demirel, S.E., & Hasan, M. M. F. (2019). Building Block-Based Synthesis and Intensification of Work-Heat Exchanger Networks (WHENS). *Processes, 7(1),* 23.

[9] Demirel, S. E., Li, J., and Hasan, M. M. F. (2019). A General Framework for Process Synthesis, Integration, and Intensification. *Industrial & Engineering Chemistry Research.* DOI: 10.1021/acs.iecr.8b05961.

576096 Hyperbolicity of Heat Transport Processes

Heat transport models are ubiquitously present in chemical, petrochemical, manufacturing and process industry as mathematical models of distributed parameter systems given by parabolic partial differential equations (PDEs). The conductive (diffusive) transport over macroscopic length scale is characterized by the well known Fourier's law and a large number of research efforts have been realized in the realm of modelling, control and optimization of such models. However, the salient characteristic of these models is that any initial disturbance in the material body is propagated instantly due to the parabolic nature of the partial differential equation (PDE) model. To eliminate this unphysical feature, we revisit the Fourier law to take into account the thermal inertia, resulting in a hyperbolic heat equation which avoids the phenomenon of infinite speed of propagation of initial data [1,2].

The main conservation laws are embedded in a modelling variety provided by the hyperbolicity of the transport systems which is physically relevant and desired property as action at distance is precluded and physically meaningful finite speed of phenomena propagation is ensured. Furthermore, the hyperbolicity mathematically ensures the well-posedness of local Cauchy problems [3]. In addition, a Stefan problem is considered, which is a specific type of heat transport boundary value problem for a partial differential equation describing the heat distribution evolution in a phase changing medium [4]. Since the moving interface is unknown a priori, the solution needs to take into account determination of the moving boundary position as well as accurate heat propagation. Finally, we provide derivation of the hyperbolic heat transport equations, the stability analysis and the numerical results for simple heat diffusion problem on fixed domain and Stefan problem.

Literature Cited

1. Cattaneo C. On a form of heat equation which eliminates the paradox of instantaneous propagation. *C R Acad Sci Paris*. 1958;pp. 431–433.

2. Vernotte P. Les paradoxes de la theorie continue de l'equation de la chaleur. *C R Acad Sci Paris*. 1958;246:3154–3155.

3. Fischer AE, Marsden JE. The Einstein evolution equations as a first-order quasi-linear symmetric hyperbolic system, I. *Communications in Mathematical Physics*. 1972;28(1):1–38.

4. Gupta S. *The Classical Stefan Problem. Basic Concepts, Modelling and Analysis*. North-Holland: Applied mathematics and Mechanics. 2003.

569568 Constructing Efficient Surrogate Models for High-Dimensional First Principles Systems Under Uncertainty

A plethora of chemical engineering processes are informed by sophisticated numerical simulations based on first-principles models (e.g., consisting of a set of coupled partial differential equations) that exhibit highly nonlinear behavior and are computationally expensive. It is often the case that the parameters in these models as well as the initial and boundary conditions in dynamical systems are uncertain and, thus, this uncertainty is propagated to the model outputs. Therefore, relying on deterministic simulation results could possibly lead to erroneous outcomes for the design, operation, and control of these processes. The computational burden associated with solving the entire set of equations can be prohibitive when one wishes to run the model for the whole spectrum of possible input parameter values, even in this era of high-performance computing, especially in the context of optimization applications. Metamodeling techniques aim at alleviating this problem by approximating first-principles model responses under uncertainty with efficient black-box mappings that can serve as surrogate models for the original high-fidelity models [1]. The primary notion in metamodeling is to calibrate the surrogate models using the fewest possible runs of the original model without compromising their predictive capability.

In this work, we explore the effectiveness of various well-established meta-modeling techniques applied to a basic, but relevant crystallization problem, i.e., the growth/dissolution of a 2D crystal population through cooling/heating cycles. This technique is used, possibly in combination with milling, to manipulate the crystal shape of active pharmaceutical ingredients, and efforts have been recently devoted to optimize and control this process. Even though relatively simple, such a problem represents an interesting test-case for metamodeling since it is highly nonlinear, inherently dynamic, and industrially relevant. [2,3] The crystallization system is described by a system of integro-partial differential equations, whose parameters are inherently uncertain. The parameter uncertainties are modeled by suitable probability distributions selected according to available experimental data and expert knowledge.

We first employ sparse polynomial chaos expansions (PCE) whose coefficients are estimated based on least-angle regression and explore their scalability with an increasing number of uncertain inputs. PCEs have been traditionally used for uncertainty quantification in numerous applications (e.g., see [4]). We then present the fundamentals of Gaussian Processes (i.e., Kriging) [5] and explore their coupling with sparse PCEs in order to create surrogate models with optimal local and global predictive capabilities [6]. Finally, the use of NARX neural networks [7] is demonstrated to deal with the continuous dynamic behavior of the system. The accuracy of these three surrogate modeling approaches in capturing the highly nonlinear spatio-temporal dynamics of the crystallization process at hand is assessed using state-of-the-art cross-validation and validation measures, and their computational aspects are discussed.

[1] Bhosekar A., Ierapetritou M., "Advances in surrogate based modeling, feasibility analysis, and optimization: A review", Computers and Chemical Engineering, 108:250-267, 2018

[2] Salvatori, F., Mazzotti, M., "Manipulation of particle morphology by crystallization, milling, and heating cycles—A mathematical modeling approach", Industrial and Engineering Chemistry Research, 56:9188–9201, 2017.

[3] Bötschi, S., Rajagopalan, A. K., Morari, M., Mazzotti, M., "Feedback control for the size and shape evolution of needle-like crystals in suspension. I. Concepts and Simulation Studies", Crystal Growth and Design, 18:4470–4483, 2018.

[4] Paulson J.A., Mesbah A., "An efficient method for stochastic optimal control with joint chance constraints for nonlinear systems", International Journal of Robust and Nonlinear Control, 1-21, 2017.

[5] Kawai S., Shimoyama K. "Kriging-model-based uncertainty quantification in computational fluid dynamics. 32nd AIAA Applied Aerodynamics Conference, 2014, 2737.

[6] Schöbi R., Sudret, B., Wiart J., "Polynomial-Chaos-based Kriging", International Journal for

Uncertainty Quantification, Begell House Publishers, 5:171-193, 2015.

[7] Mai C.V., Spiridonakos M.D., Chatzi E.N., Sudret B., "Surrogate modeling for stochastic dynamical systems by combining nonlinear autoregressive with exogenous input models and polynomial chaos expansions", International Journal for Uncertainty Quantification, 6:313-339, 2016.

574364 Additive Manufacturing of Catalyst Support Structures with in Operando Adjustable Mass Transport and Flow Characteristics

In conventional randomly packed bed reactors, the flow and mass transport characteristics are predefined by the particle diameter and the tube diameter. Structured reactors offer the opportunity to tailor mass transport and flow field according to the needs of the reaction system. In this regard, additively manufactured so-called Periodic Open Cellular Structures (POCS) as catalyst supports have shown their advantages over conventional randomly packed bed reactors in terms of heat management and pressure drop [1, 2]. However, since the internal flow field, pressure drop and therefore the resulting mass transport characteristics are mainly influenced by the geometrical properties of the unit cell, for POCS the resulting transport characteristics of the structure is already determined and fixed by the choice and design of the unit cell. To achieve more flexibility, we have recently proposed a new type of so-called interpenetrating POCS (interPOCS) which are formed by inserting a diamond unit cell based structure in the cavities of a second diamond cell structure. These interPOCS allow for in operando flow field and mass transport adjustments by shifting the (relative) position of the loose structure [3].

With detailed computational fluid dynamics (CFD) based simulations with an additional in-house particle-tracking algorithm implementation we systematically investigated the flow field and mass transport characteristics within the interPOCS in dependency of the relative positioning of the loose structure. The highly adjustable structure allows for a broad variation of mass transport characteristics in a catalytic reactor without changing the system's periphery. This functionality enables mass transport optimization for a given system and represents a completely new application of additively manufactured catalyst support structures for highly flexible in operando tuning of flow field and mass transport characteristics.

References

[1] C. Busse, H. Freund, and W. Schwieger (2018). *Chem. Eng. Process.* 124, 199-214.

[2] A. Inayat, M. Klumpp, M. Lämmermann, H. Freund and W. Schwieger (2016). *Chem. Eng. J.* 287, 704-719.

[3] G. Do, T. Stiegler, M. Fiegl, L. Adler, C. Körner, A. Bösmann, H. Freund, W. Schwieger and P. Wasserscheid (2017). *Ind. Eng. Chem. Res.* 45, 13402-13410.

561588 Role of 3D Printed and LED-Driven Photostation in Photocatalytic and Photo-Fenton Activity of Iron Oxide Doped Graphitic Carbon Nitride

The degree of photocatalysis or any photolysis reaction depends on the intensity of light. In most cases, a major portion of light is not used as it does not reach the reaction zone thereby wasting a significant portion of light and energy, as well. In order to handle this situation, we have designed and 3D printed a photostation with complex parabolic reflective surface, so that all the incident light beams can reach the reaction zone. The photostation is fitted with two chips-on-board (COB) light emitting diodes (LEDs) reaching the total (visible) light intensity within the reaction zone over 290,000 lux (For comparison, it is commonly reported that the sun emits anywhere from 32,000 to 100,000 lux onto the Earth). Light simulation analysis confirmed that all incident light beams can reach the reaction zone of photostation. In order to test the performance of this photostation, we have used 0.5 to 2 wt.% iron oxide doped graphitic carbon nitride (g-C_3N_4) to degrade a model water pollutant, dicamba (3,6-Dichloro-2-methoxybenzoic acid), which is common herbicide, in presence of traces of hydrogen peroxide and visible light. It has been confirmed experimentally that the rate of degradation of dicamba is directly proportional to the light intensity and geometry of photostation thereby confirming the positive role of photostation design. The rapid degradation of dicamba under this study can be attributed to the four distinct phenomena of photocatalytic activity of carbon nitride, quenching of electron/hole pairs and generation of additional reactive hydroxyl radicals by hydrogen peroxide, Fenton and photo-Fenton activity of iron oxide component of carbon nitride in presence of hydrogen peroxide and photocatalytic activity of iron oxide alone in conjuncture with carbon nitride. Liquid Chromatography Mass spectrometry (LCMS) analysis of the degradation products showed loss of chlorine from the aromatic ring and evidence of free radical addition reactions in the course of photocatalysis.

557053 Solution-Based 3D Printing of Hierarchical Microporous Polymers for Energy-Efficient Adsorption

Current 3D printing methods have strong limitations in the classes of polymers that can be printed. This means that many polymers of technological interest cannot currently be 3D printed. Here, we demonstrate a generalizable method for 3D printing of viscoelastic polymer solutions and apply it to both porous (a polymer of intrinsic microporosity, PIM-1) and non-porous (e.g., cellulose acetate, Matrimid) polymers. Successful ternary ink formulations that result in the creation of three-dimensional structures with hierarchical porosity require balancing of solution thermodynamics (phase separation), mass transfer (solvent evaporation), and rheology. To demonstrate the utility of this approach, a microporous polymer (PIM-1) that cannot be printed with current 3D printing technologies is 3D printed into a high-efficiency respirator exhibiting hierarchical porosity ranging from sub-nanometer to millimeter pores. These short mass transfer contactors (1.27 cm) were challenged with a flow of toluene vapor (10000 ppm) in N_2 gas. Owing to the precisely designed gas distribution channels, the 3D printed respirator were found to create a pure (< 1 ppm toluene) N_2 stream for 1.7 hours, 6 times longer than traditional structures fabricated from the same PIM-1 material, and more than 4,000 times the residence time of gas in the bed. This solution-based additive manufacturing approach significantly extends the material spectrum of 3D printing technologies. By 3D printing adsorptive materials into precisely designed mass transfer contactor, the performance, energy efficiency, and material efficiency can be significantly improved.

559790 Titanium/Titania Electrocatalyst for Hydrogen Evolution Synthesized Via a Combination of 3D Printing and Anodization

A titanium core/titania coating, hierarchically structured catalyst was fabricated via combining 3D printing and in-situ anodization. 3D printing enabled control of the macro-porosity of the titanium core. Anodization was used in order to grow highly-ordered TiO_2 nanotubes (NTs) on the metal core. The dimensions of the TiO_2 nanotubes were controlled via tuning of anodization parameters. The combination of both fabrication techniques allows for hierarchical control of the catalyst structure.

The performance of the reported 3D Ti/TiO_2 electrocatalyst was tested for hydrogen-evolution reaction (HER) via water electrolysis. Titanium dioxide is a known photo-electro-catalysts for water splitting, but competitive electrocatalytic activity for HER in the absence of light has not been reported. Onset potentials for HER as low as -45 mV vs. RHE were obtained for our electrodes in acid medium, making them competitive with current state-of-the-art electrodes aiming to replace platinum. Tafel slopes between 40 and 52 mV dec^{-1} point towards a Volmer-Heyrovsky mechanism, with the Heyrovsky reaction as possible controlling step.

The proposed electrocatalysts feature the following advantages: (1) the fabrication method for the 3D Ti/TiO_2 catalyst is simple, scalable and can be readily incorporated as printed reactor components into different technologies; (2) the 3D Ti/TiO_2 catalyst features a flow-through architecture that not only enables more efficient utilization of the surface area of the catalyst but that could enable large-scale, continuous-flow hydrogen-production reactor technologies; and (3) minimization of electrical resistance between active surface area (TiO_2 NTs) and current collector.

556480 Advanced Manufacturing Enabling Process Intensification for Catalytic Processes

Advanced manufacturing is a transformative and disruptive approach to industrial production that enables the rapid synthesis of virtually any geometry composed of a wide range of materials, such as polymers, ceramics, and metals. Process intensification is a set of novel approaches to process engineering and equipment design that leads to substantially smaller, more selective, and energy-efficient processing plants. The integration of advanced manufacturing with process intensification accelerates process development and deployment through rapid prototyping and enables new hardware that overcomes transport limitations. Advanced manufacturing (ex. 3D printing or additive manufacturing) is the physical backbone of Industry 4.0 that can take advantage of the conclusions of big data and machine learning by allowing design freedom. This expansion of the material design window can be utilized to reduce GHG intensity for novel or existing processes.

Numerous conversion technologies in the energy industry suffer from rapid coke build-up and subsequent deactivation of a catalytic material that is exposed to a carburizing environment. The time scale of deactivation often determines the reactor technology and process configuration employed. Commercial examples of deactivating catalysts range from seconds in a fluid bed catalytic cracking unit to minutes for moving bed propane dehydrogenation and days for continuous catalytic reformers or even years for steam methane reformers. A common step for these processes is the regeneration of the catalytic material via combustion of the coke. However, the cyclic process of coking in a carburizing environment followed by combustion in an oxidizing environment can often be detrimental to catalyst performance.

This work will discuss how advanced manufacturing methods can enable construction of materials capable of rapidly switching between two different environments with enhanced catalytic activity. Such materials capable of withstanding cyclic carburizing and oxidizing steps can be efficiently integrated into novel reactor concepts. The conversion of light gas is an excellent probe reaction for elucidating this behavior. In an effort to search for robust materials, honeycomb structures composed of metal were fabricated by direct metal laser sintering (DMLS), a 3D printing technique that involves the layer-wise successive deposition of micron-sized powder followed by laser induced sintering. The cyclic operation that alternates between carburizing and oxidizing environments leads to compositional modification at the surface of the monolith channels that imparts in-situ catalytic activity.

AUTHOR INDEX

AUTHOR INDEX

AUTHOR INDEX